高等院校英文教学数学系列丛书

概率论与数理统计导学(英文)

Guide to Probability and Statistics

桂文豪　张　悦　任君汝　编著

清华大学出版社
北京交通大学出版社
·北京·

内容简介

本书是根据作者多年的全英文教学经验编写而成的，是与作者编写的《概率论与数理统计（英文）》相配套的学习辅导用书。

本书主要围绕概率与随机事件、随机变量及其分布、多维随机变量及其分布、随机变量的数字特征、大数定律和中心极限定理、参数估计、假设检验、线性回归分析设计问题，并通过系统、详尽的解答分析，以及对题目背后内涵和关系的深入挖掘来帮助读者进一步提高概率论与数理统计的基本理论水平和实践应用能力。在编写过程中，作者吸取了国内外优秀教材和辅导用书的优点，注重理论与实践相结合。本书系统性强，图例丰富，突出统计思想，着力培养学生分析问题和解决实际问题的能力。

本书可用作高等院校理工科各专业本科生“概率论与数理统计”课程全英文或双语教材的辅助用书，也可供工程技术人员、科技工作者参考。

图书在版编目（CIP）数据

概率论与数理统计导学 = Guide to Probability and Statistics：英文/桂文豪，张悦，任君汝编著．—北京：北京交通大学出版社：清华大学出版社，2023. 11

ISBN 978-7-5121-5085-0

Ⅰ．①概…　Ⅱ．①桂…　②张…　③任…　Ⅲ．①概率论-高等学校-教学参考资料②数理统计-高等学校-教学参考资料　Ⅳ．①O21

中国国家版本馆 CIP 数据核字（2023）第 192323 号

概率论与数理统计导学
GAILÜLUN YU SHULI TONGJI DAOXUE

责任编辑：严慧明
出版发行：清 华 大 学 出 版 社　邮编：100084　电话：010-62776969　http：//www. tup. com. cn
　　　　　北京交通大学出版社　邮编：100044　电话：010-51686414　http：//www. bjtup. com. cn
印 刷 者：北京鑫海金澳胶印有限公司
经　　销：全国新华书店
开　　本：185 mm×260 mm　印张：13. 25　字数：299 千字
版 印 次：2023 年 11 月第 1 版　2023 年 11 月第 1 次印刷
定　　价：39. 00 元

本书如有质量问题，请向北京交通大学出版社质监组反映．对您的意见和批评，我们表示欢迎和感谢．
投诉电话：010-51686043，51686008；传真：010-62225406；E-mail：press@bjtu.edu.cn.

前　　言

为了适应信息化和全球化的社会发展趋势，培养具有国际竞争力的人才是高等教育的重要目标。全英文和双语教学是实现这一目标的有效手段，它不仅有利于学生外语水平的提高，也有利于他们专业知识的学习。“概率论与数理统计”是理工科相关专业的一门重要公共基础课，在各个领域都有其广泛的应用。

作者编写的《概率论与数理统计（英文）》自2018年出版以来，得到不少院校师生的广泛好评，许多读者来信询问是否有配套辅导书。在这一实际需求下，作者结合当前的实际教学需求，根据多年的全英文教学经验编写了本书。

本书结合全英文教学的特点，基于国际班学生的现有知识结构和学习需求，着力培养学生国际化视野，力求知识结构严谨，体系完整，紧跟统计学科发展的最新动态，便于学生进行学习。本书的主要特色有：

（1）注重归纳概率统计知识点的共性和总结概率统计规律，增加章节之间的内在联系，使读者对知识的掌握更加系统和灵活，有效提高解决实际问题的技能。

（2）在习题讲解方面，不是简单地给出答案，而是注重思考和分析过程，并对题目涉及的知识点进行认真点评，对某些题目进行多角度解析，启发和引导学生深入思考，促进学生更好地掌握解题的方法和技巧，指导学生加深对理论的理解和掌握。

（3）注重挖掘题目背后的内涵和应用，增加概率统计与微积分、几何代数、经济学、工程学等其他学科之间的横向联系，丰富学生学习的内容，提高学生解决实际问题的能力，扩大课程的影响力。

由于作者水平有限，本书肯定有不妥或不完善的地方，敬请广大读者给予批评指正，欢迎提出宝贵意见和建议。作者联系方式：桂文豪，北京市海淀区北京交通大学数学与统计学院，邮箱 whgui@ bjtu. edu. cn，邮编 100044。

作者

2023 年 5 月

目　　录

Introduction to Probability

Summary of Knowledge

Exercise Solutions

Problem 1.1 Suppose that one card is to be selected from a deck of 20 cards that contains 10 red cards numbered from 1 to 10 and 10 blue cards numbered from 1 to 10. Let A be the event that a card with an even number is selected, let B be the event that a blue card is selected, and let C be the event that a card with a number less than 5 is selected. Describe the sample space S and describe each of the following events both in words and as subsets of S:

(1) $A\cap B\cap C$; (2) $B\cap C^c$; (3) $A\cup B\cup C$; (4) $A\cap(B\cup C)$; (5) $A^c\cap B^c\cap C^c$.

Solution:

The sample space is $S=\{R_1,R_2,R_3,\cdots,R_{10},B_1,B_2,B_3,\cdots,B_{10}\}$, where R_1 means a red card numbered 1, and so on.

$A=\{R_2,R_4,R_6,R_8,R_{10},B_2,B_4,B_6,B_8,B_{10}\}$

$B=\{B_1,B_2,B_3,B_4,B_5,B_6,B_7,B_8,B_9,B_{10}\}$

$C=\{R_1,R_2,R_3,R_4,B_1,B_2,B_3,B_4\}$

(1) Blue card numbered 2 or 4. $\{B_2,B_4\}$.

(2) Blue card numbered 5, 6, 7, 8, 9, or 10. $\{B_5,B_6,B_7,B_8,B_9,B_{10}\}$.

(3) Any blue card or a red card numbered 1, 2, 3, 4, 6, 8, or 10.
$\{R_1,R_2,R_3,R_4,R_6,R_8,R_{10},B_1,B_2,B_3,B_4,B_5,B_6,B_7,B_8,B_9,B_{10}\}$.

(4) Blue card numbered 2, 4, 6, 8, or 10, or red card numbered 2 or 4. $\{B_2,B_4,B_6,B_8,B_{10},R_2,R_4\}$.

(5) Red card numbered 5, 7, or 9. $\{R_5,R_7,R_9\}$.

Problem 1.2 Two dice are rolled. Let A be the event that the sum of the two dice is odd, let B be the event that at least one of the dice has a 1 showing, and let C be the event that the sum is 5. Describe the events $A\cap B, A\cup B, B\cap C, A\cap B^c$, and $A\cap B\cap C$.

Solution:

Let x_1 be the number which appears on the first die and x_2 be the number which appears on the second die.

$$A=\{(x_1,x_2)\mid x_1+x_2=\text{odd}\}$$
$$=\{(1,2),(1,4),(1,6),(2,1),(2,3),(2,5),(3,2),(3,4),(3,6),(4,1),(4,3),(4,5),(5,2),(5,4),(5,6),(6,1),(6,3),(6,5)\}.$$

$$B=\{(1,x_2),(x_1,1)\}$$
$$=\{(1,1),(1,2),(1,3),(1,4),(1,5),(1,6),(2,1),(3,1),(4,1),(5,1),(6,1)\}.$$

$$C=\{(x_1,x_2)\mid x_1+x_2=5\}$$
$$=\{(1,4),(4,1),(2,3),(3,2)\}.$$

$$A\cap B=\{(1,2),(1,4),(1,6),(2,1),(4,1),(6,1)\}.$$

$$A\cup B=\{(1,1),(1,2),(1,3),(1,4),(1,5),(1,6),(2,1),(2,3),(2,5),(3,1),(3,2),(3,4),(3,6),(4,1),(4,3),(4,5),(5,1),(5,2),(5,4),(5,6),(6,1),(6,3),(6,5)\}.$$

$$B\cap C=\{(1,4),(4,1)\}.$$

$$A\cap B^c=\{(2,3),(2,5),(3,2),(3,4),(3,6),(4,3),(4,5),(5,2),(5,4),(5,6),(6,3),(6,5)\}.$$

$$A\cap B\cap C=\{(1,4),(4,1)\}.$$

Problem 1.3 Consider the set $S=\{1,2,3,4,5,6,7,8,9\}$ and its subsets $A=\{1,3,5,7,9\}$, $B=\{2,4,6,8\}$, $C=\{1,2,3,4\}$, $D=\{7,8\}$. Describe the following sets: (1) B^c; (2) $A\cap D$; (3) $A\cup B$; (4) $(C\cup D)^c$; (5) $A\cap(B\cup D)$.

Solution:

(1) $B^c=\{1,3,5,7,9\}=A$.

(2) $A\cap D=\{7\}$.

(3) $A\cup B=\{1,2,3,4,5,6,7,8,9\}=S$.

(4) $(C\cup D)^c=\{5,6,9\}$.

(5) $A\cap(B\cup D)=\{7\}$.

Problem 1.4 Suppose A,B,C are three events, and $P(A)=P(B)=P(C)=\frac{1}{4}$, $P(AB)=P(BC)=0$, $P(AC)=\frac{1}{8}$, determine the probability that at least one of the three events will occur.

Solution:

$ABC\subset AB$, $0\leqslant P(ABC)\leqslant P(AB)=0$, $P(ABC)=0$.

According to the addition rule for probability,

$$\begin{aligned} P(A\cup B\cup C) &= P(A)+P(B)+P(C)-P(AB)-P(AC)-P(BC)+P(ABC) \\ &= \frac{1}{4}+\frac{1}{4}+\frac{1}{4}-0-\frac{1}{8}-0+0 \\ &= \frac{5}{8}. \end{aligned}$$

Note: $P(AB)=0$ does not mean $AB=\varnothing$ and $ABC=\varnothing$.

Problem 1.5 An urn contains 5 white balls and 3 red balls. Draw 4 balls without replacement. Find the probability that 3 white balls are drawn.

Solution:

Let A denote the event that 3 white balls are drawn. The event A will occur if and only if we select 3 white balls out of the 5 white balls and 1 red ball out of the 3 red balls, therefore,

$$P(A)=\frac{C_5^3C_3^1}{C_8^4}=\frac{3}{7}\approx 0.428\ 6.$$

Problem 1.6 In a group of 20 persons, there are 5 females. If 3 persons are picked out of 20 at random, what is the probability that they all are females?

Solution:

There are C_5^3 ways to pick a group of 3 females out of the 5 females. There are C_{20}^3 ways to pick any group of 3 persons. So the probability is

$$\frac{C_5^3}{C_{20}^3}=\frac{10}{1\ 140}=\frac{1}{114}\approx 0.008\ 8.$$

Problem 1.7 Consider two events A and B with $P(A)=0.6$ and $P(B)=0.7$. Determine the maximum and minimum possible values of $P(AB)$ and the conditions under which each of these values is attained.

Solution:

(1) $AB\subset A, AB\subset B$, so $P(AB)\leqslant P(A)=0.6, P(AB)\leqslant P(B)=0.7$. The maximum possible value of $P(AB)$ is 0.6, and when $A\subset B, AB=A$, $P(AB)=P(A)=0.6$.

(2) $P(A\cup B)=P(A)+P(B)-P(AB)$, so $P(AB)=P(A)+P(B)-P(A\cup B)=1.3-P(A\cup B)$. Since $P(A\cup B)\leqslant 1, P(AB)\geqslant 0.3$. The minimum possible value of $P(AB)$ is 0.3 when $P(A\cup B)=1$.

Problem 1.8 Ten persons in a room are wearing badges marked 1 through 10. Three persons are chosen at random and the numbers of their badges are recorded. Find the probability:

(1) the smallest number is 5; (2) the largest number is 5.

 **Solution**:

(1) 5 must be chosen, the other two numbers are selected from 6~10,

$$P=\frac{1\times C_5^2}{C_{10}^3}=\frac{1}{12}\approx 0.083\ 3.$$

(2) 5 must be chosen, the other two numbers are selected from 1~4,

$$P=\frac{1\times C_4^2}{C_{10}^3}=\frac{1}{20}=0.05.$$

Problem 1.9 Choose randomly an integer from 1~2 000. Find the probability:

(1) the selected number is divisible by 6; (2) the selected number is divisible by 8; (3) the selected number is divisible by 6 and 8.

 **Solution**:

(1) There are $\left[\frac{2\ 000}{6}\right]=333$ multiples of 6 in that range. So the probability that the selected number is divisible by 6 is $\frac{333}{2\ 000}$. Here $[x]$ denotes the greatest integer less than or equal to x.

(2) Similarly, there are $\left[\frac{2\ 000}{8}\right]=250$ multiples of 8 in that range. So the probability that the selected number is divisible by 8 is $\frac{250}{2\ 000}=\frac{1}{8}$.

(3) The least common multiple of 6 and 8 is 24. There are $\left[\frac{2\ 000}{24}\right]=83$ multiples of 24 in that range. So the probability that the selected number is divisible by 6 and 8 is $\frac{83}{2\ 000}$.

Problem 1.10 A closet contains n pairs of shoes. If $2r$ shoes are chosen at random, (where $2r<n$), what is the probability that (1) the chosen shoes will contain no matching pair? (2) there will be exactly one complete pair?

 **Solution**:

(1) There are C_{2n}^{2r} possible outcomes. To have no pairs, choose $2r$ pairs from n pairs, then choose one of the two shoes from each pair, the probability is

$$P=\frac{C_n^{2r}2^{2r}}{C_{2n}^{2r}}.$$

Consider it in another way, for $2n$ shoes of n pairs, there are P_{2n}^{2r} ways to select $2r$ shoes in order. The unpaired way is to take one first with $2n$ ways, the second cannot take the first pair, so there are only $2n-2$ ways, analogously, the third one has $2n-4$ ways, $\cdots$, the $2r$-th shoe has

$2n-2(2r-1)$ ways, then the probability is

$$P=\frac{2n\times(2n-2)\times\cdots\times[2n-2(2r-1)]}{P_{2n}^{2r}}=\frac{C_n^{2r}2^{2r}}{C_{2n}^{2r}}.$$

(2) First choose one pair from n pairs of shoes, next choose $2r-2$ pairs from the remaining $n-1$ pairs, then take one shoe from each of the $2r-2$ pairs, so the probability is

$$P=\frac{C_n^1C_{n-1}^{2r-2}2^{2r-2}}{C_{2n}^{2r}}.$$

Problem 1.11 If event A is drawing a queen from a deck of cards and event B is drawing a king from the remaining cards, are events A and B dependent or independent?

Solution:

$P(A)=\frac{4}{52}=\frac{1}{13}$. Let C denote the event that a king is drawn at the first time to draw a card. By the total probability rule,

$$P(B)=P(C)P(B\mid C)+P(C^c)P(B\mid C^c)=\frac{4}{52}\times\frac{3}{51}+\frac{48}{52}\times\frac{4}{51}=\frac{1}{13}.$$

So, $P(A)P(B)=\frac{1}{13}\times\frac{1}{13}=\frac{1}{169}$. However, $P(AB)=\frac{4}{52}\times\frac{4}{51}=\frac{4}{663}$. $P(AB)\neq P(A)P(B)$, therefore, the events A and B are dependent.

Problem 1.12 Bag A contains 9 red marbles and 3 green marbles. Bag B contains 9 black marbles and 6 orange marbles. Find the probability of selecting a green marble from bag A and a black marble from bag B.

Solution:

The probability of drawing a green marble from bag A is the number of green marbles divided by the total number of marbles, and is $\frac{3}{12}=\frac{1}{4}$. The probability of drawing a black marble from bag B is $\frac{9}{15}=\frac{3}{5}$. Since these are independent events, the probability is the product of the two probabilities: $\frac{1}{4}\times\frac{3}{5}=\frac{3}{20}$.

Problem 1.13 Consider two events A and B such that $P(A)=\frac{1}{3}$ and $P(B)=\frac{1}{2}$. Determine the value of $P(B\cap A^c)$ for each of the following conditions: (1) A and B are disjoint; (2) $A\subset B$; (3) $P(A\cap B)=\frac{1}{8}$.

Solution:

(1) If A and B are disjoint, then $B \subset A^c$ and $BA^c = B$, so $P(BA^c) = P(B) = \frac{1}{2}$.

(2) If $A \subset B$, then $B = A \cup (BA^c)$, where A and BA^c are disjoint. So $P(B) = P(A) + P(BA^c)$. That is, $\frac{1}{2} = \frac{1}{3} + P(BA^c)$, so $P(BA^c) = \frac{1}{6}$.

(3) $B = (BA) \cup (BA^c)$. Also, BA and BA^c are disjoint, so, $P(B) = P(BA) + P(BA^c)$, that is, $\frac{1}{2} = \frac{1}{8} + P(BA^c)$, so $P(BA^c) = \frac{3}{8}$.

Problem 1.14 If 50 percent of the families in a certain city subscribe to the morning newspaper, 65 percent of the families subscribe to the afternoon newspaper, and 85 percent of the families subscribe to at least one of the two newspapers, what percentage of the families subscribe to both newspapers?

Solution:

Let A be the event that a randomly selected family subscribes to the morning paper, and let B be the event that a randomly selected family subscribes to the afternoon paper. We are told that $P(A) = 0.5$, $P(B) = 0.65$, and $P(A \cup B) = 0.85$. We are asked to find $P(A \cap B)$. Using the addition rule,

$$P(A \cup B) = P(A) + P(B) - P(A \cap B).$$

Therefore,

$$P(A \cap B) = P(A) + P(B) - P(A \cup B) = 0.5 + 0.65 - 0.85 = 0.3.$$

Problem 1.15 Suppose that A, B, and C are three independent events such that $P(A) = \frac{1}{4}$, $P(B) = \frac{1}{3}$, and $P(C) = \frac{1}{2}$.

(1) Determine the probability that none of these three events will occur.

(2) Determine the probability that exactly one of these three events will occur.

Solution:

(1) $P(A^c \cap B^c \cap C^c) = P(A^c)P(B^c)P(C^c) = \frac{3}{4} \times \frac{2}{3} \times \frac{1}{2} = \frac{1}{4}$.

(2) The desired probability is $P(A \cap B^c \cap C^c) + P(A^c \cap B \cap C^c) + P(A^c \cap B^c \cap C) = \frac{1}{4} \times \frac{2}{3} \times \frac{1}{2} + \frac{3}{4} \times \frac{1}{3} \times \frac{1}{2} + \frac{3}{4} \times \frac{2}{3} \times \frac{1}{2} = \frac{11}{24}$.

Problem 1.16 A box contains 100 items of which 4 are defective. Two items are chosen at random from the box. What is the probability of selecting

(1) 2 defectives if the first item is not replaced;

(2) 2 defectives if the first item is put back before choosing the second item;

(3) 1 defective and 1 non-defective if the first item is not replaced?

Solution:

(1) On the first draw, there are 4 defectives in the box out of the 100 total items. If we have already chosen one of the defectives on the first draw, then on the second draw, there will be 3 defectives left out of the 99 items in the box. The required probability is:

$$\frac{4}{100}\times\frac{3}{99}=\frac{1}{825}\approx 0.001\ 2.$$

(2) Both the first draw and the second draw have the same probability of getting a defective, i.e. 4 in 100. The required probability is:

$$\frac{4}{100}\times\frac{4}{100}=\frac{1}{625}=0.001\ 6.$$

(3) We can either:

i. Get a defective item on the first draw (4 chances in 100) then a non-defective on the second draw (96 non-defectives out of 99 items left in the box); OR

ii. Get a non-defective item first (96 chances in 100) then a defective item (4 in the remaining 99 items). So the probability is

$$\frac{4}{100}\times\frac{96}{99}+\frac{96}{100}\times\frac{4}{99}\approx 0.077\ 6.$$

Problem 1.17 The probability that a student passes Mathematics test is $\frac{2}{3}$ and the probability that he passes English test is $\frac{4}{9}$. If the probability that he will pass at least one subject test is $\frac{4}{5}$, what is the probability that he will pass both subject tests?

Solution:

Let A denote the event that the student passes Mathematics test, and let B be the event that he passes English test.

$$P(A)=\frac{2}{3}, P(B)=\frac{4}{9}, P(A\cup B)=\frac{4}{5}.$$

Therefore, the desired probability is

$$P(AB)=P(A)+P(B)-P(A\cup B)=\frac{2}{3}+\frac{4}{9}-\frac{4}{5}=\frac{14}{45}.$$

Problem 1.18 When two balanced dice are rolled, what is the probability that the sum of the numbers on the dice is 6 or 9?

Solution:

There are 36 possible outcomes in the sample space. Let A be the event that the sum of the numbers on the dice is 6, and let B be the event that the sum of the numbers on the dice is 9.

$$A=\{(1,5),(2,4),(3,3),(4,2),(5,1)\},P(A)=\frac{5}{36}$$

$$B=\{(3,6),(4,5),(5,4),(6,3)\},P(B)=\frac{4}{36}$$

Since A and B are disjoint, the probability that the sum of the numbers on the dice is 6 or 9 is

$$P(A\cup B)=P(A)+P(B)=\frac{9}{36}=\frac{1}{4}.$$

Problem 1.19 An elevator has 4 passengers and stops at 7 floors. It is equally likely that a person will get off at any one of the 7 floors. Find the probability that no 2 or more passengers leave at the same floor.

Solution:

There are 7^4 possible outcomes in the sample space. If the 4 passengers are to get off at different floors, the first passenger can get off at any one of the seven floors, the second passenger can then get off at any one of the other six floors, etc. Thus, the probability is

$$\frac{7\times6\times5\times4}{7^4}=\frac{120}{343}\approx0.349\ 9.$$

Problem 1.20 The elevator of a four-floor building leaves the first floor with six passengers and stops at all of the remaining three floors. If it is equally likely that a passenger gets off at any of these three floors, what is the probability that at each stop of the elevator at least one passenger departs?

Solution 1:

We define the events E,A,B,C as follows:

E=at least one passenger departs at each stop;

A=at least one passenger gets off at the second floor;

B = at least one passenger gets off at the third floor;

C = at least one passenger gets off at the fourth floor,

then the probability of at least one passenger departing at each floor is:

$$P(E)=P(A\cap B\cap C)=1-P((A\cap B\cap C)^c)=1-P(A^c\cup B^c\cup C^c).$$

Now $P(A^c\cup B^c\cup C^c)$ can be expanded as:

$$P(A^c\cup B^c\cup C^c)=P(A^c)+P(B^c)+P(C^c)-P(A^c\cap B^c)-P(A^c\cap C^c)-\\P(B^c\cap C^c)+P(A^c\cap B^c\cap C^c).$$

Let p be the probability that a passenger gets off at a particular floor, then $p=\frac{1}{3}$. $P(A^c\cap B^c\cap C^c)$ is the probability that none of the passengers depart at any of thc floors which is zero. $P(A^c)$ is the probability that none of the passengers get off on the second floor, which is $(1-p)^6$. Similarly, $P(B^c)=(1-p)^6$ and $P(C^c)=(1-p)^6$. $P(A^c\cap B^c)$ is the probability that none of the passengers gets off on the second or third floor, i.e. all the passengers get off on the fourth floor, thus $P(A^c\cap B^c)=p^6$. Similarly, $P(A^c\cap C^c)=p^6$ and $P(B^c\cap C^c)=p^6$. Finally,

$$\begin{aligned}P(E)&=1-3(1-p)^6+3p^6\\&=1-3\times\left(\frac{2}{3}\right)^6+3\times\left(\frac{1}{3}\right)^6\\&=1-\frac{7}{27}=\frac{20}{27}\\&\approx 0.740\ 7.\end{aligned}$$

Solution 2:

There are $3^6=729$ possible outcomes in the sample space. The event that at least one passenger leaves each floor includes the following cases and corresponding combinations,

(i) 4-1-1: $3\times C_6^4\times C_2^1\times C_1^1=90$ ways;

(ii) 3-2-1: $6\times C_6^3\times C_3^2\times C_1^1=360$ ways;

(iii) 2-2-2: $C_6^2\times C_4^2\times C_2^2=90$ ways;

The total number of outcomes in the event is 540. The probability is $\frac{540}{729}=\frac{20}{27}\approx 0.740\ 7$.

Problem 1.21 Six fair dice are tossed. What is the probability that at least two of them show the same face?

Solution:

Determine first the probability that the six numbers are all different. With six rolls, there are $6^6=46\ 656$ possible outcomes. The outcomes with all different rolls are the permutations of six distinct items. There are $6!=720$ outcomes, so this probability is $\frac{720}{46\ 656}=0.015\ 4$, the desired probability is 1−0.015 4=0.984 6.

Problem 1.22 If 10 balls are thrown at random into 20 boxes, what is the probability that no box will receive more than one ball?

Solution:

There are 20^{10} possible outcomes in the sample space. If the 10 balls are to be thrown into different boxes, the first ball can be thrown into any one of the 20 boxes, the second ball can then be thrown into any one of the other 19 boxes, etc. Thus, there are $20\times19\times18\times\cdots\times11$ possible outcomes in the event. So the probability is $\frac{20!}{10!\times20^{10}}=0.065\ 5$.

Problem 1.23 Suppose we pick three people at random. For each of the following questions, ignore the special case where someone might be born on February 29th, and assume that births are evenly distributed throughout the year.

(1) What is the probability that the first two people share a birthday?

(2) What is the probability that at least two people share a birthday?

Solution:

(1) Since there are 365 possible birthdays for each of 3 people, the sample space S will contain 365^3 outcomes, all of which will be equally probable. Since the first person's birthday could be any one of the 365 days, the second person's birthday is same with the first one, and the third person's birthday could then be any of the other 364 days, therefore, the probability that the first two people share a birthday is

$$\frac{365\times1\times364}{365^3}\approx0.002\ 7.$$

(2) The number of outcomes in A for which all 3 birthdays will be different is P_{365}^3, since the first person's birthday could be any one of the 365 days, the second person's birthday could then be any of the other 364 days, and so on. Hence, the probability that all 3 persons will have different birthdays is

$$P(A)=\frac{P_{365}^3}{365^3}\approx0.991\ 8.$$

The probability $P(A^c)$ that at least two of the people will have the same birthday is therefore $P(A^c)=1-P(A)=1-0.991\ 8=0.008\ 2$.

Problem 1.24 A box contains 24 light bulbs, of which four are defective. If a person selects four bulbs from the box at random, without replacement, what is the probability that all four bulbs will be defective?

Solution:

The sample space consists of all subsets(unordered) of 4 bulbs drawn from the 24 bulbs in the box. There are $C_{24}^{4}=10\ 626$ such subsets. There is only one subset that has all four defectives, so the probability we want is $\frac{1}{10\ 626}$.

Problem 1.25 A box contains 24 light bulbs of which four are defective. If one person selects 10 bulbs from the box in a random manner, and a second person then takes the remaining 14 bulbs, what is the probability that all four defective bulbs will be obtained by the same person?

Solution:

Let A be the event that all four defective bulbs will be obtained by the first person, and let B be the event that all four defective bulbs will be obtained by the second person. A and B are disjoint.

$$P(A)=\frac{C_4^4 C_{20}^6}{C_{24}^{10}}\approx 0.019\ 76$$

$$P(B)=\frac{C_4^0 C_{20}^{10}}{C_{24}^{10}}\approx 0.094\ 20$$

$$P(A\cup B)=P(A)+P(B)\approx 0.114\ 0.$$

Problem 1.26 A deck of 52 cards contains four aces. If the cards are shuffled and distributed in a random manner to four players so that each player receives 13 cards, what is the probability that all four aces will be received by the same player?

Solution 1:

Call the four players A, B, C, and D. Here we view the problem as that of distributing 52 cards into four groups(group 1 for A, group 2 for B, and so on), each with 13 cards. There are

$$C_{52}^{13}C_{39}^{13}C_{26}^{13}C_{13}^{13}=\frac{52!}{13!\times 13!\times 13!\times 13!}$$

ways of doing this.

Now if A receives all 4 aces, then the remaining $52-4=48$ cards are to be distributed among A, B, C, D, so that A gets $13-4=9$ cards and each of B, C, and D gets 13 cards. There are

$$C_{48}^{9}C_{39}^{13}C_{26}^{13}C_{13}^{13}=\frac{48!}{9!\times 13!\times 13!\times 13!}$$

ways of doing this. Similarly, B can receive all 4 aces in

$$C_{48}^{13}C_{35}^{9}C_{26}^{13}C_{13}^{13}=\frac{48!}{9!\times 13!\times 13!\times 13!}$$

ways, and so on. Hence, total number of ways of distributing 52 cards into four players, each receiving 13 cards, in such a manner that all four aces are received by the same player is

$$4\times\frac{48!}{9!\times 13!\times 13!\times 13!}$$

Therefore, the required probability is

$$P=\frac{4\times\dfrac{48!}{9!\times 13!\times 13!\times 13!}}{\dfrac{52!}{13!\times 13!\times 13!\times 13!}}\approx 0.010\,6.$$

Solution 2:

Note that we can also view the given problem as that of choosing positions for the four aces in the deck. Each card occupies a single position.

Out of the 52 available positions we need to choose 4 for placing the 4 aces. This can be done in C_{52}^4 ways. Now choose one player out of four, in $C_4^1=4$ ways. In order for that (chosen) player to receive all four aces, the aces must be placed among the 13 cards received by that player, which can be done in C_{13}^4 ways. Hence, total number ways of choosing the positions so that all four aces are received by the same player is $4\,C_{13}^4$. Therefore, the required probability is $\dfrac{4\,C_{13}^4}{C_{52}^4}\approx 0.010\,6$.

Solution 3:

We can also view the given problem as that of choosing positions for the 52 cards in the deck. The sample space contains 52! outcomes.

4 ways to arrange the four aces to one of the four persons, and there are P_{13}^4 ways to choose the four positions for the four aces for the chosen person, there are 48! ways to arrange the remaining 48 cards in the 48 remaining positions.

$$P=\frac{4\times P_{13}^4\times 48!}{52!}=0.010\,6.$$

Problem 1.27 Suppose that five guests check their hats when they arrive at a restaurant, and that these hats are returned to them in a random order when they leave. Determine the probability that

(1) no guests will receive the proper hat.

(2) at least two of the guests receive their own hats.

Solution:

(1) Let A_i be the event that guest i receives his own hat ($i=1,\cdots,n$), we shall determine the value, for $n\geqslant 2$,

$$\begin{aligned}P(n)&=P(A_1^c\cap A_2^c\cap\cdots\cap A_n^c)\\&=P\left[\left(\bigcup_{i=1}^{n}A_i\right)^c\right]\\&=1-P\left(\bigcup_{i=1}^{n}A_i\right)\\&=1-\left[1-\frac{1}{2!}+\frac{1}{3!}-\frac{1}{4!}+\cdots+(-1)^{n+1}\frac{1}{n!}\right]\\&=\frac{1}{2!}-\frac{1}{3!}+\frac{1}{4!}-\cdots+(-1)^n\frac{1}{n!}.\end{aligned}$$

Then, $n=5$, the probability that no guests will receive the proper hat is

$$P(5)=\frac{1}{2!}-\frac{1}{3!}+\frac{1}{4!}-\frac{1}{5!}=\frac{11}{30}=0.366\ 7.$$

(2) Let X be the number of guests who receive their own hats. $X=0,1,2,3,4,5$. We shall determine the value of $P(X\geqslant 2)$. Determine first the probability $P(X=1)$. $X=1$ means that one guest receives his own hat, but no other guests receive the proper hats.

$$\begin{aligned}P(X=1)&=C_5^1\times\frac{1}{5}\times P(4)\\&=\frac{1}{2!}-\frac{1}{3!}+\frac{1}{4!}\\&=\frac{3}{8}.\end{aligned}$$

$$P(X=0)=P(5)=\frac{1}{2!}-\frac{1}{3!}+\frac{1}{4!}-\frac{1}{5!}=\frac{11}{30}.$$

Therefore, $P(X\geqslant 2)=1-P(X=0)-P(X=1)=1-\frac{11}{30}-\frac{3}{8}=\frac{31}{120}\approx 0.258\ 3.$

Problem 1.28 A box contains 4 red balls and 8 blue balls. One ball is selected at random and its color is observed. The ball is then returned to the box and one additional ball of the same color is also put into the box. A second ball is then selected at random, its color is observed, and it is returned to the box together with one additional ball of the same color. Each time another ball is selected, the process is repeated. If four balls are selected, what is the probability that the first three balls will be red and the fouth ball will be blue?

 **Solution:**

Let R_1 be the event that the first ball is red, let R_2 be the event that the second ball is red, let R_3 be the event that the third ball is red, and let B_4 be the event that the fourth ball is blue.

By the multiplication rule,

$$\begin{aligned}P(R_1R_2R_3B_4)&=P(R_1)P(R_2\mid R_1)P(R_3\mid R_1R_2)P(B_4\mid R_1R_2R_3)\\&=\frac{4}{12}\times\frac{5}{13}\times\frac{6}{14}\times\frac{8}{15}\\&=\frac{8}{273}.\end{aligned}$$

Problem 1.29 A box contains three cards. One card is red on both sides, one card is green on both sides, and one card is red on one side and green on the other. One card is selected from the box at random, and the color on one side is observed. If this side is green, what is the probability that the other side of the card is also green?

Solution:

Let the cards be labelled as C_1, C_2, C_3 with C_1 being red on both sides, C_2 being green on both sides and C_3 being red on one side and green on the other. Also, let G be the event that the observed side is green. Then,

$$P(G)=P(C_1)P(G \mid C_1)+P(C_2)P(G \mid C_2)+P(C_3)P(G \mid C_3)$$
$$=\frac{1}{3}\times 0+\frac{1}{3}\times 1+\frac{1}{3}\times\frac{1}{2}$$
$$=\frac{1}{2}.$$

Therefore,

$$P(C_2 \mid G)=\frac{P(C_2)P(G \mid C_2)}{P(G)}=\frac{\frac{1}{3}\times 1}{\frac{1}{2}}=\frac{2}{3}.$$

Problem 1.30 What is the probability that the total of two dice will be greater than 8, given that the first die is a 6?

Solution:

Let A be the event that the first die is a 6, and let B be the event that the total of two dice will be greater than 8.

$$P(A)=\frac{1}{6}, A\cap B=\{(6,3),(6,4),(6,5),(6,6)\}, P(A\cap B)=\frac{4}{36}=\frac{1}{9}.$$

Hence, the probability that the total of two dice will be greater than 8, given that the first die is a 6 is

$$P(B \mid A)=\frac{P(AB)}{P(A)}=\frac{2}{3}.$$

Problem 1.31 Given that 5% of men and 0.25% of women are color-blind. A person is randomly selected. Assume males and females to be in equal numbers.

(1) What is the probability that the selected person is color-blind?

(2) If the selected person is color-blind, what is the probability that he is male?

Solution:

Let A_1, A_2 denote the events that the selected person is male or female, respectively, and let B be the event that the person is color-blind.

(1) By the law of total probability,

$$P(B)=P(A_1)P(B\mid A_1)+P(A_2)P(B\mid A_2)=\frac{1}{2}\times 0.05+\frac{1}{2}\times 0.002\ 5=0.026\ 25.$$

(2) By the Bayes' formula,

$$P(A_1\mid B)=\frac{P(A_1)P(B\mid A_1)}{P(B)}=\frac{\frac{1}{2}\times 0.05}{0.026\ 25}\approx 0.952\ 4.$$

Problem 1.32 A machine produces defective parts with three different probabilities depending on its state of repair. If the machine is in good working order, it produces defective parts with probability 0.02. If it is wearing down, it produces defective parts with probability 0.1. If it needs maintenance, it produces defective parts with probability 0.3. The probability that the machine is in good working order is 0.8, the probability that it is wearing down is 0.1, and the probability that it needs maintenance is 0.1. Compute the probability that a randomly selected part will be defective.

Solution:

Let B_1 be the event that the machine is in good working order. Let B_2 be the event that the machine is wearing down. Let B_3 be the event that it needs maintenance. Let A be the event that a selected part is defective.

It is given that $P(B_1)=0.8$ and $P(B_2)=P(B_3)=0.1$. $P(A\mid B_1)=0.02, P(A\mid B_2)=0.1$, and $P(A\mid B_3)=0.3$.

By the law of total probability,

$$\begin{aligned}P(A)&=\sum_{i=1}^{3}P(B_i)P(A\mid B_i)\\&=0.8\times 0.02+0.1\times 0.1+0.1\times 0.3\\&=0.056.\end{aligned}$$

Problem 1.33 Suppose that a medical test has a 92% chance of detecting a disease if the person has it and a 94% chance of correctly indicating that the disease is absent if the person really does not have the disease. Suppose that 10% of the population has the disease. (1) What is the probability that a randomly chosen person will test positive? (2) What is the probability of having the disease given a positive test?

Solution:

Let B denote the event that the person has the disease, and let A be the event that he tests positive. $P(A \mid B)=0.92, P(A^c \mid B^c)=0.94, P(B)=0.1, P(B^c)=0.9$.

(1) By the law of total probability,

$$P(A)=P(B)P(A \mid B)+P(B^c)P(A \mid B^c)=0.1\times0.92+0.9\times(1-0.94)=0.146.$$

(2) By the Bayes' formula,

$$P(B \mid A)=\frac{P(B)P(A \mid B)}{P(A)}=\frac{0.1\times0.92}{0.146}\approx0.63.$$

Problem 1.34 Suppose that A and B are events such that $P(A)=0.3, P(B)=0.4, P(A \mid B)=0.5$. Evaluate

(1) $P(AB), P(A\cup B)$.

(2) $P(B \mid A)$.

(3) $P(B \mid A\cup B)$.

(4) $P(A^c\cup B^c \mid A\cup B)$.

**Solution:**

(1) $P(AB)=P(B)P(A \mid B)=0.4\times0.5=0.2$,

$$P(A\cup B)=P(A)+P(B)-P(AB)=0.3+0.4-0.2=0.5.$$

(2) $P(B \mid A)=\dfrac{P(AB)}{P(A)}=\dfrac{0.2}{0.3}=\dfrac{2}{3}$.

(3) $P(B \mid A\cup B)=\dfrac{P[B\cap(A\cup B)]}{P(A\cup B)}=\dfrac{P(B)}{P(A\cup B)}=\dfrac{0.4}{0.5}=\dfrac{4}{5}$.

(4)
$$\begin{aligned}P(A^c\cup B^c \mid A\cup B)&=\frac{P[(A^c\cup B^c)\cap(A\cup B)]}{P(A\cup B)}\\&=\frac{P[(AB)^c\cap(A\cup B)]}{P(A\cup B)}\\&=\frac{P(A\cup B)-P(AB)}{P(A\cup B)}\\&=\frac{0.5-0.2}{0.5}\\&=\frac{3}{5}.\end{aligned}$$

Problem 1.35 Suppose that A, B, and C are events such that A and B are independent, $P(A\cap B\cap C)=0.04, P(C \mid A\cap B)=0.25$, and $P(B)=4P(A)$. Evaluate $P(A\cup B)$.

Solution:

$$P(A\cap B)=\frac{P(A\cap B\cap C)}{P(C\mid A\cap B)}=\frac{0.04}{0.25}=0.16.$$

On the other hand, by independence,

$$P(A\cap B)=P(A)P(B)=4[P(A)]^2.$$

Hence, $4[P(A)]^2=0.16, P(A)=0.2$. It follows that

$$P(A\cup B)=P(A)+P(B)-P(A\cap B)=0.2+4\times0.2-0.16=0.84.$$

Problem 1.36 Suppose $0<P(A)<1, 0<P(B)<1, P(A\mid B)+P(A^c\mid B^c)=1$. Are events A and B independent?

Solution 1:

Since $P(A\mid B^c)+P(A^c\mid B^c)=1, P(A\mid B)=P(A\mid B^c)$,

$$\frac{P(AB)}{P(B)}=\frac{P(AB^c)}{P(B^c)}=\frac{P(AB)+P(AB^c)}{P(B)+P(B^c)}=P(A)$$

$$P(AB)=P(A)P(B).$$

Therefore, A and B are independent.

Solution 2:

$$\begin{aligned}P(A\mid B)+P(A^c\mid B^c)&=\frac{P(AB)}{P(B)}+\frac{P(A^cB^c)}{P(B^c)}\\&=\frac{P(AB)}{P(B)}+\frac{1-P(A\cup B)}{1-P(B)}\\&=\frac{P(AB)}{P(B)}+\frac{1-P(A)-P(B)+P(AB)}{1-P(B)}\\&=\frac{P(AB)[1-P(B)]+P(B)[1-P(A)-P(B)+P(AB)]}{P(B)[1-P(B)]}\\&=\frac{P(AB)-P(A)P(B)+P(B)-[P(B)]^2}{P(B)[1-P(B)]}\\&=\frac{P(AB)-P(A)P(B)}{P(B)[1-P(B)]}+1.\end{aligned}$$

Thus, $P(AB)-P(A)P(B)=0, P(AB)=P(A)P(B)$, A and B are independent.

Problem 1.37 Suppose that 13 cards are selected at random from a regular deck of 52 playing cards.

(1) If it is known that at least one ace has been selected, what is the probability that at least two aces have been selected?

(2) If it is known that the ace of hearts has been selected, what is the probability that at least two aces have been selected?

Solution:

(1) Let X denote the number of aces selected. Then

$$P(X=i)=\frac{C_4^i C_{48}^{13-i}}{C_{52}^{13}}, i=0,1,2,3,4$$

$$\begin{aligned} P(X\geqslant 2 \mid X\geqslant 1) &= \frac{P(X\geqslant 2)}{P(X\geqslant 1)} \\ &= \frac{1-P(X=0)-P(X=1)}{1-P(X=0)} \\ &= \frac{1-0.303\ 8-0.438\ 8}{1-0.303\ 8} \\ &= 0.369\ 7. \end{aligned}$$

(2) Let H denote the event that the ace of hearts is obtained and let B denote the event that at least two aces have been selected. The required probability is $P(B \mid H)=\frac{P(HB)}{P(H)}$. We know

$$P(H)=\frac{1\times C_{51}^{12}}{C_{52}^{13}}=0.25.$$

Method 1: Let A denote the event that the ace of hearts and no other aces are obtained. Then $A\subset H$, and A^c denotes the event that there is no ace of hearts or there exist other aces. HA^c denotes the event that the ace of hearts and other aces are obtained, which is exactly the event HB.

$$\begin{aligned} P(HB) &= P(HA^c) \\ &= P(H)-P(A) \\ &= 0.25-\frac{1\times C_{48}^{12}}{C_{52}^{13}} \\ &= 0.25-0.109\ 7 \\ &= 0.140\ 3. \end{aligned}$$

Therefore, $P(B \mid H)=\frac{P(HB)}{P(H)}=\frac{0.140\ 3}{0.25}=0.561\ 2.$

Method 2: Let $A_i (i=1,2,3)$ denote the event that the ace of hearts and exactly i other aces are obtained. Then

$$\begin{aligned} P(HB) &= P(A_1)+P(A_2)+P(A_3) \\ &= \frac{1\times C_3^1\times C_{48}^{11}}{C_{52}^{13}}+\frac{1\times C_3^2\times C_{48}^{10}}{C_{52}^{13}}+\frac{1\times C_3^3\times C_{48}^{9}}{C_{52}^{13}} \\ &= 0.140\ 3. \end{aligned}$$

Therefore, $P(B \mid H)=\frac{P(HB)}{P(H)}=\frac{0.140\ 3}{0.25}=0.561\ 2.$

Method 3: If it is known that the ace of hearts has been selected, the sample space is reduced and only 51 cards are left. In this case, we need find the probability that at least one ace is obtained when selecting 12 cards.

$$P=1-\frac{C_{48}^{12}}{C_{51}^{12}}=0.561\ 2.$$

Problem 1.38 A biased coin (with probability of obtaining a head equal to $p>0$) is tossed repeatedly and independently until the first head is observed. Compute the probability that the first head appears at an even numbered toss.

Solution 1:

Let H_1 denote a head on the first toss, and let E be the event that the first head on even numbered toss. Sample space Ω consists of all possible infinite binary sequences of coin tosses.

By the law of total probability,

$$P(E)=P(H_1)P(E\mid H_1)+P(H_1^c)P(E\mid H_1^c).$$

Now, clearly, $P(E\mid H_1)=0$ (given H_1 that a head appears on the first toss, E cannot occur) and also $P(E\mid H_1^c)$ can be seen to be given by

$$P(E\mid H_1^c)=P(E^c)=1-P(E).$$

(given that a head does not appear on the first toss, the required conditional probability is merely the probability that the sequence concludes after a further odd number of tosses, that is, the probability of E^c). Hence $P(E)$ satisfies

$$P(E)=p\times 0+(1-p)\times(1-P(E)).$$

Thus, $P(E)=\frac{1-p}{2-p}$.

Solution 2:

Consider the partition of E into $E_1, E_2, \cdots$, where E_k is the event that the first head occurs on the $2k$-th toss. Then

$$E=\bigcup_{k=1}^{\infty}E_k, P(E)=P\left(\bigcup_{k=1}^{\infty}E_k\right)=\sum_{k=1}^{\infty}P(E_k).$$

Now $P(E_k)=(1-p)^{2k-1}p$ (that is, $2k-1$ tails, then a head), so

$$\begin{aligned}P(E)&=\sum_{k=1}^{\infty}(1-p)^{2k-1}p\\&=\frac{p}{1-p}\sum_{k=1}^{\infty}(1-p)^{2k}\\&=\frac{p}{1-p}\times\frac{(1-p)^2}{1-(1-p)^2}\\&=\frac{1-p}{2-p}.\end{aligned}$$

Problem 1.39 An engineering company advertises a job in three papers A, B and C. It is known that these papers attract undergraduate engineering readerships in the proportions 2 : 3 : 1. The probabilities that an engineering undergraduate sees and replies to the job advertisement in these papers are 0.002, 0.001, and 0.005, respectively. Assume that the undergraduate sees only one job advertisement.

(1) If the engineering company receives only one reply to its advertisements, calculate the probability that the applicant has seen the job advertised in paper A.

(2) If the company receives two replies, what is the probability that both applicants saw the job advertised in paper A?

Solution:

Let A, B, C denote the events that the person is a reader of paper A, B, C, respectively, and let D be the event that the reader applies for the job.

$$P(A)=\frac{2}{2+3+1}=\frac{1}{3}, P(B)=\frac{1}{2}, P(C)=\frac{1}{6}.$$

(1) By the Bayes' formula,

$$P(A \mid D)=\frac{P(D \mid A)P(A)}{P(D \mid A)P(A)+P(D \mid B)P(B)+P(D \mid C)P(C)}$$

$$=\frac{0.002\times\frac{1}{3}}{0.002\times\frac{1}{3}+0.001\times\frac{1}{2}+0.005\times\frac{1}{6}}$$

$$=\frac{1}{3}.$$

(2) Assume that the replies and readerships are independent, so

$$P(\text{both applicants saw the job in paper } A)=P(A \mid D)\times P(A \mid D)=\left(\frac{1}{3}\right)^2=\frac{1}{9}.$$

Problem 1.40 Two players A and B are competing at a trivia quiz game involving a series of questions. On any individual question, the probabilities that A and B give the correct answer are p_A and p_B, respectively, for all questions, with outcomes for different questions being independent. The game finishes when a player wins by answering a question correctly. Compute the probability that A wins if

(1) A answers the first question;

(2) B answers the first question.

Solution:

Define:

- sample space Ω——all possible infinite sequences of answers;
- event A——A answers the first question;
- event F——game ends after the first question;
- event W——A wins.

We want to find $P(W|A)$ and $P(W|A^c)$. Using the law of total probability, and the partition given by $\{F,F^c\}$

$$P(W|A)=P(W|A\cap F)P(F|A)+P(W|A\cap F^c)P(F^c|A).$$

Now, clearly

$P(F|A)=P[$A answers the first question correctly$]=p_A, P(F^c|A)=1-p_A$, and $P(W|A\cap F)=1$, but $P(W|A\cap F^c)=P(W|A^c)$, so that

$$P(W|A)=1\times p_A+P(W|A^c)\times(1-p_A)=p_A+P(W|A^c)(1-p_A). \qquad (*)$$

Similarly,

$$P(W|A^c)=P(W|A^c\cap F)P(F|A^c)+P(W|A^c\cap F^c)P(F^c|A^c).$$

We have

$P(F|A^c)=P[$B answers the first question correctly$]=p_B, P(F^c|A^c)=1-p_B$, but $P(W|A^c\cap F)=0$. Finally $P(W|A^c\cap F^c)=P(W|A)$, so that

$$P(W|A^c)=0\times p_B+P(W|A)\times(1-p_B)=P(W|A)(1-p_B). \qquad (**)$$

Solving $(*)$ and $(**)$ simultaneously gives

$$P(W|A)=\frac{p_A}{1-(1-p_A)(1-p_B)}, P(W|A^c)=\frac{(1-p_B)p_A}{1-(1-p_A)(1-p_B)}.$$

Random Variables and Distributions

Summary of Knowledge

Exercise Solutions

Problem 2.1 An urn contains 4 balls marked with numbers 0, 1, 1, 2, respectively. Draw 2 balls with replacement. Let X be the product of the numbers of the balls selected. Find the PF of X.

 **Solution**:

$\{0,1,1,2\}\times\{0,1,1,2\}=\{0,1,2,4\}$.

$P(X=0)=P(\text{at least one ball marked with } 0)=\frac{1}{4}+\frac{1}{4}-\frac{1}{4}\times\frac{1}{4}=\frac{7}{16}$.

$P(X=1)=P(2 \text{ balls marked with number } 1)=\frac{1}{2}\times\frac{1}{2}=\frac{1}{4}$.

$P(X=2)=P(\text{one ball marked with } 1\text{, the other ball marked with } 2)=\frac{2}{4}\times\frac{1}{4}+\frac{1}{4}\times\frac{2}{4}=\frac{1}{4}$.

$P(X=4)=P(2 \text{ balls marked with number } 2)=\frac{1}{4}\times\frac{1}{4}=\frac{1}{16}$.

The PF of X is given as follows:

X	0	1	2	4
$P(X=x)$	$\frac{7}{16}$	$\frac{1}{4}$	$\frac{1}{4}$	$\frac{1}{16}$

Problem 2.2 Suppose that a random variable X has a discrete distribution with the following PF:

$$f(x)=\begin{cases} cx, & x=1,2,\cdots,5 \\ 0, & \text{otherwise} \end{cases}.$$

Determine the value of the constant c.

Solution:

The sum of all the individual probabilities $f(x)$ must be equal to 1. Since $1 = \sum_{x=1}^{5} f(x) = 15c$, we have $c = \frac{1}{15}$.

Problem 2.3 Suppose that a random variable X has a discrete distribution with the following PF:

$$f(x) = \begin{cases} \frac{c}{2^x}, & x = 0, 1, 2, \cdots \\ 0, & \text{otherwise} \end{cases}.$$

Find the value of the constant c.

Solution:

The sum of all the individual probabilities $f(x)$ must be equal to 1. $\sum_{x=0}^{\infty} f(x) = 1$, it implies that

$$c = \frac{1}{\sum_{x=0}^{\infty} 2^{-x}} = \frac{1}{2}.$$

In the above expression we applied the formula for a geometric series

$$1 + r + r^2 + r^3 + \cdots = \frac{1}{1-r} \text{ for } |r| < 1.$$

Problem 2.4 Suppose that two balanced dice are rolled, and let X denote the absolute value of the difference between the two numbers that appear. Determine and sketch the PF of X.

Solution:

The sample space is shown as follows:

(1,1),(1,2),(1,3),(1,4),(1,5),(1,6)
(2,1),(2,2),(2,3),(2,4),(2,5),(2,6)
(3,1),(3,2),(3,3),(3,4),(3,5),(3,6)
(4,1),(4,2),(4,3),(4,4),(4,5),(4,6)
(5,1),(5,2),(5,3),(5,4),(5,5),(5,6)
(6,1),(6,2),(6,3),(6,4),(6,5),(6,6)

Where the first entry in the pair (m, n) indicates the number on the first die, and the second entry indicates the number on the second die. Since the dice are balanced, the outcomes are equally likely. Thus, all the outcomes have probability $\frac{1}{36}$.

The possible values for X are 0, 1, 2, 3, 4, 5. The outcomes that yield $X=0$ lie along the main diagonal and there are six of them. It follows that $P(X=0)=\frac{6}{36}=\frac{1}{6}$. Similarly, we find that there are 10 outcomes that yield $X=1$; 8 outcomes for $X=2$; 6 outcomes for $X=3$; 4 outcomes for $X=4$, and 2 outcomes for $X=5$.

Hence, the PF $f(x)$ is as follows (see Figure 2.1).

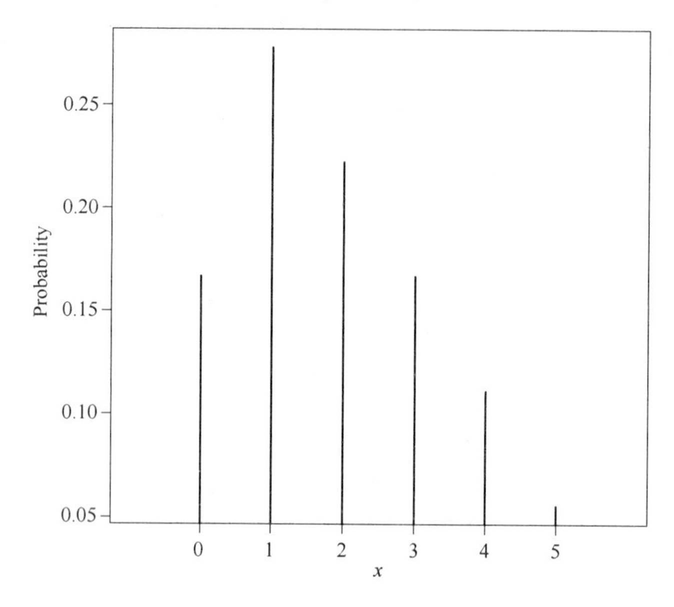

Figure 2.1

X	0	1	2	3	4	5
$P(X=x)$	$\frac{1}{6}$	$\frac{5}{18}$	$\frac{2}{9}$	$\frac{1}{6}$	$\frac{1}{9}$	$\frac{1}{18}$

Problem 2.5 A civil engineer is studying a left-turn lane that is long enough to hold seven cars. Let X be the number of cars in the lane at the end of a randomly chosen red light. The engineer believes that the probability that $X=x$ is proportional to $(x+1)(8-x)$ for $x=0,\cdots,7$ (the possible values of X).

(1) Find the PF of X.

(2) Find the probability that X will be at least 5.

Solution:

(1) The PF of X is $f(x)=c(x+1)(8-x)$ for $x=0,\cdots,7$, where c is chosen so that $\sum_{x=0}^{7} f(x)=1$. So,

c is one over $\sum_{x=0}^{7}(x+1)(8-x)$. The sum equals 120, so $c=\frac{1}{120}$. The PF of X is as follows

$$f(x)=\frac{(x+1)(8-x)}{120} \text{ for } x=0,\cdots,7.$$

(2)

$$\begin{aligned}P(X\geqslant 5)&=P(X=5)+P(X=6)+P(X=7)\\&=\frac{(5+1)\times(8-5)}{120}+\frac{(6+1)\times(8-6)}{120}+\frac{(7+1)\times(8-7)}{120}\\&=\frac{1}{3}.\end{aligned}$$

Problem 2.6 Suppose a discrete random variable X has the PF given by

X	3	5	7	8	9	10	12
$P(X=x)$	0.08	0.10	0.16	0.25	0.20	0.03	0.18

What is the probability that: (1) $X>9$; (2) $4<X<11$; (3) X is 7.

Solution:

(1) $P(X>9)=P(X=10)+P(X=12)=0.03+0.18=0.21$.

(2) $P(4<X<11)=P(X=5)+P(X=7)+P(X=8)+P(X=9)+P(X=10)=0.10+0.16+0.25+0.20+0.03=0.74$.

OR $P(4<X<11)=1-P(X=3)-P(X=12)=1-0.08-0.18=0.74$.

(3) $P(X \text{ is } 7)=0.16$.

Problem 2.7 Suppose that a box contains seven red balls and three blue balls. If five balls are selected at random, without replacement. Determine the PF of the number of red balls that will be obtained.

Solution:

Let X denote the number of red balls in the five selected ones. Then X is discrete and takes on the possible values 2, 3, 4 and 5. The PF of X is

$$P(X=x)=\frac{C_7^x C_3^{5-x}}{C_{10}^5}, x=2,3,4,5.$$

Problem 2.8 Suppose that a random variable X has the binomial distribution with parameters $n=15$ and $p=0.5$. Find $P(X<6)$.

Solution:

$$P(X<6)=\sum_{i=0}^{5}P(X=i)=\sum_{i=0}^{5}C_{15}^{i}\,0.5^{i}(1-0.5)^{15-i}=0.150\,9.$$

Problem 2.9 If 10 percent of the balls in a certain box are red, and if 20 balls are selected from the box at random, with replacement. What is the probability that more than three red balls will be obtained?

Solution:

Let X be the number of red balls selected from the box. X follows the binomial distribution with parameters $n=20$ and $p=0.1$.

$$\begin{aligned} P(X>3) &= 1-P(X=0)-P(X=1)-P(X=2)-P(X=3) \\ &= 1-C_{20}^{0}\times 0.1^{0}\times 0.9^{20}-C_{20}^{1}\times 0.1^{1}\times 0.9^{19}-C_{20}^{2}\times 0.1^{2}\times 0.9^{18}- \\ &\quad C_{20}^{3}\times 0.1^{3}\times 0.9^{17} \\ &= 0.133\ 0. \end{aligned}$$

Problem 2.10 The probability of winning at a certain game is 0.3. If you play the game 10 times, what is the probability that you win at most once?

Solution:

Let X denote the number of times you win, $X\sim B(10,0.3)$.

$$\begin{aligned} P(X\leqslant 1) &= P(X=0)+P(X=1) \\ &= C_{10}^{0}\times 0.3^{0}\times 0.7^{10}+C_{10}^{1}\times 0.3^{1}\times 0.7^{9} \\ &= 0.149\ 3. \end{aligned}$$

Problem 2.11 Suppose that X follows the Poisson distribution, $P(X=1)=P(X=2)$. Find $P(X=4)$.

Solution:

Since X follows the Poisson distribution $P(\lambda)$, $\lambda>0$,

$$P(X=1)=\frac{\lambda^{1}}{1!}e^{-\lambda}=P(X=2)=\frac{\lambda^{2}}{2!}e^{-\lambda}$$

Solve the equation, then $\lambda=2$. Therefore,

$$P(X=4)=\frac{\lambda^{4}}{4!}e^{-\lambda}=\frac{2^{4}}{4!}e^{-2}=0.090\ 2.$$

Problem 2.12 The number of calls coming per minute into a hotel reservation center is Poisson random variable with $\lambda=3$.

(1) Find the probability that no calls come in a given one-minute period.

(2) Assume that the number of calls arriving in two different minutes are independent. Find the probability that at least two calls will arrive in a given two-minute period.

Solution:

(1) Let X denote the number of calls coming in a given one-minute period. $X \sim P(3)$,

$$P(X=0)=\frac{3^0}{0!}e^{-3}=e^{-3}=0.049\ 8.$$

(2) Let X_1 and X_2 be the numbers of calls coming in the first and second minute, respectively. $X_1 \sim P(3)$, $X_2 \sim P(3)$, X_1 and X_2 are independent.

$$\begin{aligned}P(X_1+X_2 \geqslant 2) &= 1-P(X_1+X_2<2)\\ &= 1-[P(X_1=0,X_2=0)+P(X_1=1,X_2=0)+P(X_1=0,X_2=1)]\\ &= 1-[P(X_1=0)P(X_2=0)+P(X_1=1)P(X_2=0)+P(X_1=0)P(X_2=1)]\\ &= 1-[e^{-3}e^{-3}+3\,e^{-3}e^{-3}+3\,e^{-3}e^{-3}]\\ &= 1-7\,e^{-6}\\ &= 0.982\ 6.\end{aligned}$$

Problem 2.13 Suppose that a part has a one in a hundred chance of failing and we sample from 100 independent parts. What is the probability that none of the 100 parts fails?

Solution:

Let X denote the number of failed parts, then $X \sim B(100, 0.01)$. So the probability that none of the 100 parts fail is

$$P(X=0)=(1-0.01)^{100}=0.99^{100}\approx 0.366\ 0.$$

Or we can apply the Poisson approximation to the binomial probability. $\lambda = np = 100 \times 0.01 = 1$. $X \sim P(1)$.

$$P(X=0)=\frac{1^0}{0!}e^{-1}=e^{-1}\approx 0.367\ 9.$$

Problem 2.14 The number of E-mails received in a day has a Poisson distribution with parameter λ. On that day, the probability of one E-mail being spammed is denoted by p. Show that the number of spam E-mails received that day also follows a Poisson distribution.

Solution:

Let X denote the number of E-mails received in a day, and let Y denote the number of spam E-mails received on that day. The random variable Y takes the values of $0, 1, 2, \cdots$

$$\begin{aligned}P(Y=k) &= \sum_{n=k}^{\infty} P(X=n)P(Y=k \mid X=n)\\ &= \sum_{n=k}^{\infty} \frac{\lambda^n e^{-\lambda}}{n!} \times C_n^k p^k (1-p)^{n-k}\\ &= \sum_{n=k}^{\infty} \frac{\lambda^n e^{-\lambda}}{n!} \times \frac{n!}{k!(n-k)!} p^k (1-p)^{n-k}\end{aligned}$$

$$= \frac{(\lambda p)^k e^{-\lambda}}{k!} \sum_{n=k}^{\infty} \frac{[\lambda(1-p)]^{n-k}}{(n-k)!} \quad (\text{let } t = n-k)$$

$$= \frac{(\lambda p)^k e^{-\lambda}}{k!} \sum_{t=0}^{\infty} \frac{[\lambda(1-p)]^t}{t!}$$

$$= \frac{(\lambda p)^k e^{-\lambda}}{k!} e^{\lambda(1-p)}$$

$$= \frac{(\lambda p)^k e^{-\lambda p}}{k!},$$

for $k=0,1,2,\cdots$.

Therefore, the number of spam E - mails received on that day also follows a Poisson distribution with parameter λp.

Problem 2.15 A crate contains 50 light bulbs of which 5 are defective and 45 are not. A quality control inspector randomly samples 4 bulbs without replacement. Let X be the number of defective bulbs selected. Find the probability function of the discrete random variable X.

Solution:

This is a hyper-geometric distribution problem. The PF of X is

$$P(X=i)=\frac{C_5^i C_{45}^{4-i}}{C_{50}^4}, \quad \text{for } i=0,1,2,3,4.$$

Problem 2.16 The products produced by a machine have a 5% defective rate.

(1) What is the probability that the first defective occurs in the sixth item inspected?

(2) What is the probability that the first defective occurs in the first six inspections?

Solution:

Let X denote the number of trials until the first defective occurs. X is a geometric random variable. $X \sim \text{Geo}(0.05)$.

(1) $P(X=6)=(1-0.05)^5 \times 0.05=0.038\,7$.

(2) $P(X \leqslant 6) = \sum_{i=1}^{6}(1-0.05)^{i-1} \times 0.05=1-(1-0.05)^6=0.264\,9$.

Problem 2.17 An oil company conducts a geological study that indicates that an exploratory oil well should have a 30% chance of striking oil.

(1) What is the probability that the first strike comes on the third well drilled?

(2) What is the probability that the third strike comes on the seventh well drilled?

Solution:

(1) Let X be the number of wells drilled until the first strike of oil. To find the requested

probability, we need to find $P(X=3)$. Note that X is technically a geometric random variable, since we are only looking for one success.

$$P(X=3)=(1-0.3)^2\times 0.3=0.147.$$

(2) Let Y be the number of wells drilled until the third strike of oil. X is a negative binomial random variable with $r=3$ and $p=0.3$.

$$P(X=7)=C_{7-1}^{3-1}(1-p)^{7-3}p^3=C_6^2\times(1-0.3)^4\times 0.3^3=0.097\ 2.$$

Problem 2.18 Suppose that the PDF of a random variable X is as follows:

$$f(x)=\begin{cases}\frac{1}{36}(9-x^2), & -3\leqslant x\leqslant 3\\ 0, & \text{otherwise}\end{cases}.$$

Sketch this PDF and determine the values of the following probabilities: (1) $P(X<0)$; (2) $P(-1\leqslant X\leqslant 1)$; (3) $P(X>2)$.

Solution:

The PDF is skeched in Figure 2.2.

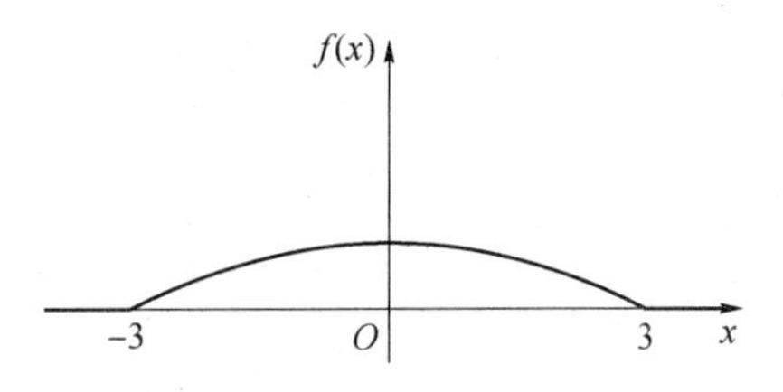

Figure 2.2

(1) $P(X<0)=\int_{-3}^{0}\frac{1}{36}(9-x^2)\,dx=0.5.$

(2) $P(-1\leqslant X\leqslant 1)=\int_{-1}^{1}\frac{1}{36}(9-x^2)\,dx=0.481\ 5.$

(3) $P(X>2)=\int_{2}^{3}\frac{1}{36}(9-x^2)\,dx=0.074\ 1.$

Problem 2.19 Suppose that the PDF of a random variable X is as follows:

$$f(x)=\begin{cases}cx^2, & 1\leqslant x\leqslant 2\\ 0, & \text{otherwise}\end{cases}.$$

(1) Find the value of the constant c and sketch the PDF.

(2) Find the values of $P\left(X>\frac{3}{2}\right)$ and $P\left(X=\frac{3}{2}\right)$.

Solution:

(1) We integrate $f(x)$ with respect to x, set the result equal to 1 and solve for c.

$$\int_{-\infty}^{\infty} f(x)\,dx = \int_{1}^{2} c\,x^2\,dx = \frac{7}{3}c = 1.$$

Therefore, $c=\frac{3}{7}$.

$$f(x)=\begin{cases}\frac{3}{7}x^2, & 1\leqslant x\leqslant 2\\ 0, & \text{otherwise}\end{cases}.$$

The PDF is plotted in Figure 2.3.

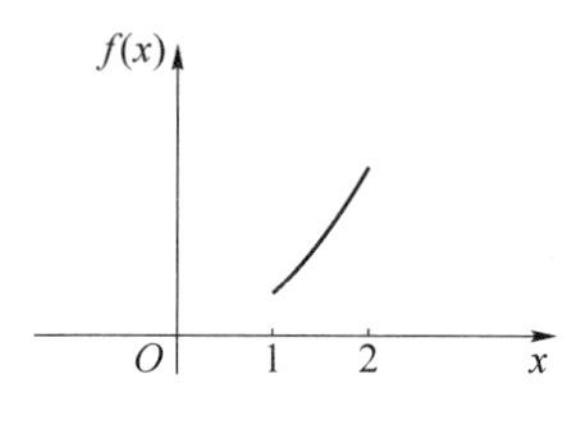

Figure 2.3

(2) $P\left(X > \frac{3}{2}\right) = \int_{\frac{3}{2}}^{+\infty} f(x)\,dx = \int_{\frac{3}{2}}^{2} \frac{3}{7}x^2\,dx = \frac{37}{56}, P\left(X = \frac{3}{2}\right) = 0.$

Problem 2.20 Suppose that the PDF of a random variable X is as follows:

$$f(x)=\begin{cases}\frac{1}{8}x, & 0\leqslant x\leqslant 4\\ 0, & \text{otherwise}\end{cases}.$$

(1) Find the value of a such that $P(X\leqslant a)=1/4$.

(2) Find the value of b such that $P(X\geqslant b)=1/2$.

Solution:

The PDF is plotted in Figure 2.4.

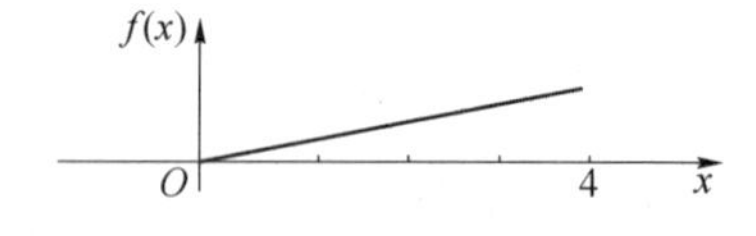

Figure 2.4

(1) $\int_0^a \frac{1}{8}x\,dx = \frac{1}{4}$, or $\frac{a^2}{16}=\frac{1}{4}$. Hence, $a=2$.

(2) $\int_b^4 \frac{1}{8}x\,dx = \frac{1}{2}$, or $1-\frac{b^2}{16}=\frac{1}{2}$. Hence, $b=2\sqrt{2}$.

Problem 2.21 Suppose that the PDF of a random variable X is as follows:

$$f(x)=\begin{cases} c\,\mathrm{e}^{-2x}, & x>0 \\ 0, & \text{otherwise} \end{cases}.$$

(1) Find the value of the constant c and sketch the PDF.

(2) Find the value of $P(X<2)$.

Solution:

(1) We integrate $f(x)$ with respect to x, set the result equal to 1 and solve for c.

$$\int_{-\infty}^{\infty} f(x)\,\mathrm{d}x = \int_{0}^{+\infty} c\,\mathrm{e}^{-2x}\,\mathrm{d}x = \frac{1}{2}c = 1.$$

Therefore, $c=2$.

$$f(x)=\begin{cases} 2\,\mathrm{e}^{-2x}, & x>0 \\ 0, & \text{otherwise} \end{cases}.$$

The PDF is plotted in Figure 2.5.

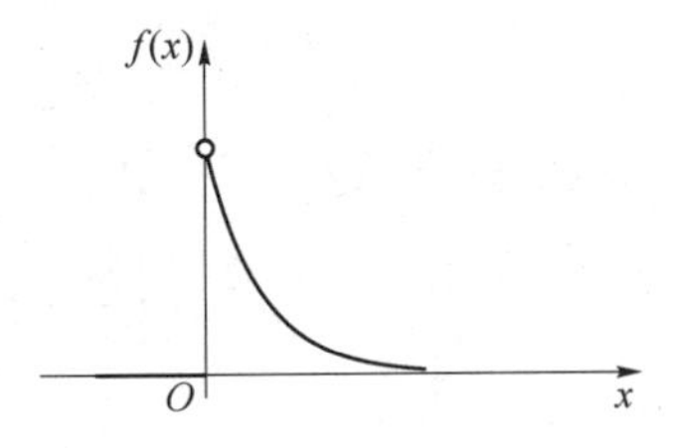

Figure 2.5

(2) $P(X<2)=\int_{-\infty}^{2} f(x)\,\mathrm{d}x = \int_{0}^{2} 2\,\mathrm{e}^{-2x}\,\mathrm{d}x = 1-\mathrm{e}^{-4} = 0.981\,7.$

Problem 2.22 Suppose that a random variable X can take only the values $-2, 0, 1$, and 4, and that the probabilities of these values are as follows: $P(X=-2)=0.4$, $P(X=0)=0.1$, $P(X=1)=0.3$, and $P(X=4)=0.2$. Sketch the CDF of X.

Solution:

(1) For $x<-2$, $F(x)=P(\varnothing)=0$.

(2) For $-2\leqslant x<0$, $F(x)=P(X\leqslant x)=P(X=-2)=0.4$.

(3) For $0\leqslant x<1$, $F(x)=P(X\leqslant x)=P(X=-2)+P(X=0)=0.5$.

(4) For $1\leqslant x<4$, $F(x)=P(X\leqslant x)=P(X=-2)+P(X=0)+P(X=1)=0.8$.

(5) For $x\geqslant 4$, $F(x)=P(X=-2)+P(X=0)+P(X=1)+P(X=4)=1$.

In summary, $F(x)=\begin{cases}0, & x<-2\\ 0.4, & -2\leq x<0\\ 0.5, & 0\leq x<1\\ 0.8, & 1\leq x<4\\ 1, & x\geq 4\end{cases}$, as shown in Figure 2.6.

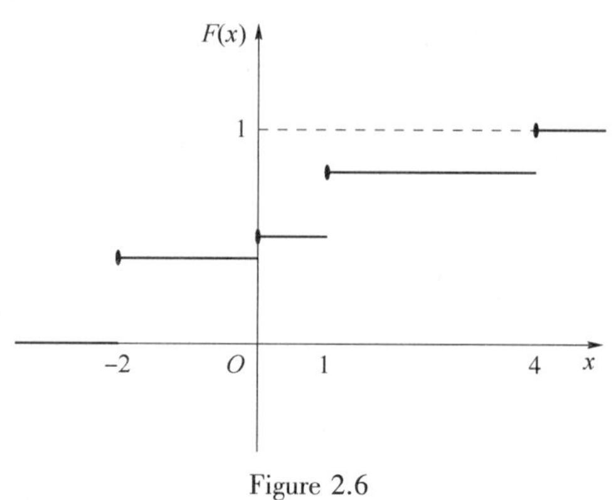

Figure 2.6

Problem 2.23 Suppose that the CDF of a random variable X is as follows:

$$F(x)=\begin{cases}0, & x\leq 0\\ \frac{1}{9}x^2, & 0<x<3\\ 1, & x\geq 3\end{cases}.$$

Find and sketch the PDF of X.

Solution:

The derivative of $F(x)$ with respect to x is the PDF $f(x)$, as shown in Figure 2.7.

$$f(x)=\begin{cases}\frac{2}{9}x, & 0<x<3\\ 0, & \text{otherwise}\end{cases}.$$

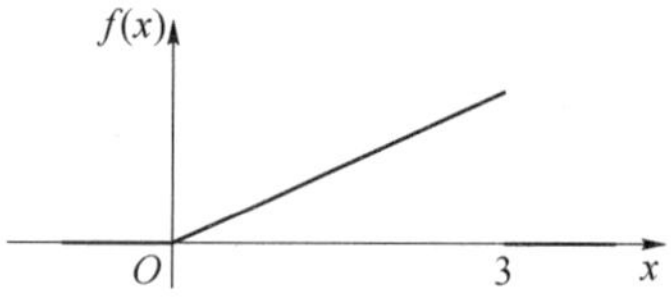

Figure 2.7

Problem 2.24 Suppose that X has the PDF

$$f(x)=\begin{cases}2x, & 0<x<1\\ 0, & \text{otherwise}\end{cases}.$$

Find and sketch the CDF of X.

Solution:

The CDF of X is $F(x)=\int_{-\infty}^{x} f(t)\,\mathrm{d}t$. Here $f(t)=\begin{cases}2t, & 0<t<1\\ 0, & \text{otherwise}\end{cases}$.

If $x\leqslant 0$, $F(x)=\int_{-\infty}^{x} 0\mathrm{d}t = 0$.

If $0<x<1$, $F(x)=\int_{0}^{x} 2t\mathrm{d}t = x^2$.

If $x\leqslant 1$, $F(x)=\int_{0}^{1} 2t\mathrm{d}t = 1$.

As shown in Figure 2.8, the CDF $F(x)$ is $F(x)=\begin{cases}0, & x\leqslant 0\\ x^2, & 0<x<1\\ 1, & x\geqslant 1\end{cases}$.

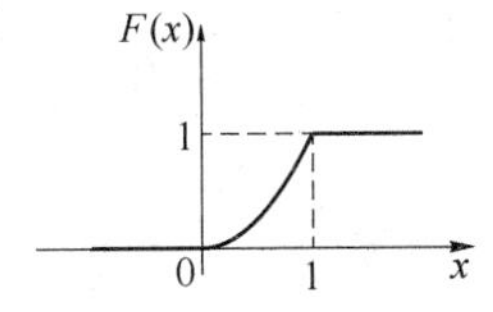

Figure 2.8

Problem 2.25 Suppose that K follows the uniform distribution over the interval $(0,5)$. Find the probability that the following quadratic equation has real roots.

$$4x^2+4Kx+K+2=0.$$

Solution:

The PDF of K is

$$f(x)=\begin{cases}\frac{1}{5}, & 0<x<5\\ 0, & \text{otherwise}\end{cases}.$$

Let A be the event that the quadratic equation has real roots, then

$$\begin{aligned}P(A)&=P[(4K)^2-4\times4\times(K+2)\geqslant 0]\\ &=P[(K+1)(K-2)\geqslant 0]\\ &=P(K\leqslant -1 \text{ or } K\geqslant 2)\\ &=P(K\leqslant -1)+P(K\geqslant 2)\\ &=0+\int_{2}^{5}\frac{1}{5}\mathrm{d}x\\ &=\frac{3}{5}.\end{aligned}$$

Problem 2.26 Suppose that X follows a discrete distribution with the PF

X	-2	-1	0	1	2	4
P	0.2	0.1	0.3	0.1	0.2	0.1

Find the probability that the following quadratic equation has real roots.

$$3a^2+2aX+X+1=0.$$

Solution:

Let A be the event that the quadratic equation has real roots, then

$$\begin{aligned}P(A)&=P[(2X)^2-4\times3\times(X+1)\geqslant0]\\&=P(X^2-3X-3\geqslant0)\\&=P\left(X\leqslant\frac{3-\sqrt{21}}{2}\text{ or }X\geqslant\frac{3+\sqrt{21}}{2}\right)\\&=P\left(X\leqslant\frac{3-\sqrt{21}}{2}\right)+P\left(X\geqslant\frac{3+\sqrt{21}}{2}\right)\\&=P(X=-2)+P(X=-1)+P(X=4)\\&=0.4.\end{aligned}$$

Problem 2.27 Suppose that the amount of waiting time X (unit: minute) a customer spends in a bank is exponentially distributed with the following density

$$f(x)=\begin{cases}\dfrac{1}{5}e^{-\frac{x}{5}}, & x>0\\0, & x\leqslant0\end{cases}.$$

The customer decides to leave if the waiting time is not less than 10 minutes. He usually visits the bank 5 times per month. Let Y be the times he is not served within a certain month. Find $P(Y\geqslant1)$.

Solution:

Y follows the binomial distribution $B(5,p)$, where

$$p=P(X>10)=\int_{10}^{+\infty}\frac{1}{5}e^{-\frac{x}{5}}dx=-e^{-\frac{x}{5}}\Big|_{10}^{+\infty}=e^{-2}.$$

Therefore,

$$P(Y\geqslant1)=1-P(Y=0)=1-(1-e^{-2})^5\approx0.516\ 7.$$

Problem 2.28 Suppose that $X\sim N(160,\sigma^2)$. If $P(120<X<200)\geqslant0.80$, what is the largest value σ can take?

Solution:

$Z=\dfrac{X-160}{\sigma}\sim N(0,1).$

$$P(120<X<200)=P\left(\frac{120-160}{\sigma}<\frac{X-160}{\sigma}<\frac{200-160}{\sigma}\right)$$
$$=P\left(-\frac{40}{\sigma}<Z<\frac{40}{\sigma}\right)$$
$$=2\Phi\left(\frac{40}{\sigma}\right)-1.$$

$2\Phi\left(\dfrac{40}{\sigma}\right)-1\geqslant 0.8, \Phi\left(\dfrac{40}{\sigma}\right)\geqslant 0.9$, using the standard normal table, we can find that $\Phi(1.28)$ $=0.9$, therefore, $\dfrac{40}{\sigma}\geqslant \Phi^{-1}(0.9)=1.28, \sigma\leqslant 31.25$.

Problem 2.29 Suppose X, the grade on a midterm exam, is normally distributed with mean 70 and standard deviation 10. The instructor wants to give 15% of the class an A. What cutoff should the instructor use to determine who gets an A?

Solution:

$X\sim N(70,10^2), Z=\dfrac{X-70}{10}\sim N(0,1).$

Suppose the cutoff should be a, we shall determine a such that $P(X>a)=0.15$.

$$P(X\leqslant a)=P\left(Z\leqslant\frac{a-70}{10}\right)=\Phi\left(\frac{a-70}{10}\right)=1-0.15=0.85.$$

Using the standard normal table, we can find that $\Phi(1.04)=0.85$, so $\dfrac{a-70}{10}=1.04$. The cutoff is $a=80.4$.

Problem 2.30 A random variable X has a normal distribution with $\sigma=10$. If the probability that the random variable will take on a value less than 82.5 is 0.821 2, what is the probability that it will take a value greater than 58.3?

Solution:

$X\sim N(\mu,10^2), Z=\dfrac{X-\mu}{10}\sim N(0,1).$

$$P(X<82.5)=P\left(Z<\frac{82.5-\mu}{10}\right)=\Phi\left(\frac{82.5-\mu}{10}\right)=0.821\ 2.$$

Using the standard normal table, we find that $\Phi(0.92)=0.821\ 2$, therefore, $\dfrac{82.5-\mu}{10}=0.92$, $\mu=73.3$.

$$P(X>58.3)=P\left(Z>\frac{58.3-73.3}{10}\right)$$
$$=P(Z>-1.50)$$
$$=P(Z<1.50)$$
$$=\Phi(1.50)$$
$$=0.933\ 2.$$

Problem 2.31 An electronic device has probabilities of 0.1, 0.001, 0.2 of failing subject to voltage of less than 220 V, 220~240 V, larger than 240 V, respectively. Suppose the voltage $X\sim N(220,25^2)$.

(1) Find the probability that the electronic device will fail.

(2) If it failed, what is the probability that it had received the voltage of 220~240 V?

 **Solution**:

$X\sim N(220,25^2)$, $Z=\frac{X-220}{25}\sim N(0,1)$.

Let A denote the event that the electronic device fails and $B_i(i=1,2,3)$ denote the voltage of less than 220 V, 220~240 V, larger than 240 V, respectively.

$P(B_1)=P(X<220)=P\left(Z<\frac{220-220}{25}\right)=P(Z<0)=0.5.$

$P(B_2)=P(220\leqslant X\leqslant 240)=P(0\leqslant Z\leqslant 0.8)=0.288\ 1.$

$P(B_3)=P(X>240)=P\left(X>\frac{240-220}{25}\right)=P(Z>0.8)=0.211\ 9.$

(1) By the law of total probability,

$$\sum_{i=1}^{3}P(B_i)P(A\mid B_i)=0.5\times 0.1+0.288\ 1\times 0.001+0.211\ 9\times 0.2=0.092\ 7.$$

(2) By the Bayes' formula,

$$P(B_2\mid A)=\frac{P(B_2)P(A\mid B_2)}{P(A)}=\frac{0.288\ 1\times 0.001}{0.092\ 7}=0.003\ 1.$$

Problem 2.32 A certain type of electronic component has a lifetime X (in hours) with probability density function given by

$$f(x)=\begin{cases}\frac{c}{x^2}, & x>1\ 000\\ 0, & x\leqslant 1\ 000\end{cases}.$$

(1) Find the constant c.

(2) Find $P(X\leqslant 1\ 800\mid X>1\ 500)$.

(3) Consider 3 independent electronic components of this type. What is the probability that only one component failed for the first 1 500 hours?

Solution:

(1) $1=\int_{-\infty}^{\infty}f(x)\,dx=\int_{1\,000}^{\infty}\frac{c}{x^2}dx=-\frac{c}{x}\Big|_{1\,000}^{\infty}=\frac{c}{1\,000}$, $c=1\,000$. The PDF of X is

$$f(x)=\begin{cases}\frac{1\,000}{x^2}, & x>1\,000\\ 0, & x\leqslant 1\,000\end{cases}.$$

(2)
$$P(X\leqslant 1\,800\mid X>1\,500)=\frac{P(1\,500<X\leqslant 1\,800)}{P(X>1\,500)}$$
$$=\frac{\int_{1\,500}^{1\,800}\frac{1\,000}{x^2}dx}{\int_{1\,500}^{+\infty}\frac{1\,000}{x^2}dx}$$
$$=\frac{\frac{1}{15}-\frac{1}{18}}{\frac{1}{15}-0}$$
$$=\frac{1}{6}.$$

(3) Let Y denote the number of components failed for the first 1 500 hours, then Y follows the binomial distribution $B(3,p)$, where

$$p=P(X<1\,500)=\int_{1\,000}^{1\,500}\frac{1\,000}{x^2}dx=\frac{1}{3}$$
$$P(Y=1)=C_3^1\left(\frac{1}{3}\right)^1\left(1-\frac{1}{3}\right)^2=\frac{4}{9}.$$

Problem 2.33 Suppose the PDF of X is

$$f(x)=\begin{cases}Ax, & 1<x<2\\ B, & 2<x<3\\ 0, & \text{otherwise}\end{cases}.$$

And $P(1<X<2)=2P(2<X<3)$. Find:

(1) the constants A and B;

(2) the CDF of X.

Solution:

(1) $1=\int_{-\infty}^{\infty}f(x)\,dx=\int_1^2 Ax\,dx+\int_2^3 B\,dx=\frac{3}{2}A+B$

$\int_1^2 Ax\,dx=2\int_2^3 B\,dx$, $\frac{3}{2}A=2B$

Solve the two equations above, we can obtain $A=\frac{4}{9}, B=\frac{1}{3}$. Therefore,

$$f(x)=\begin{cases}\frac{4}{9}x, & 1<x<2\\ \frac{1}{3}, & 2<x<3\\ 0, & \text{otherwise}\end{cases}.$$

(2) The CDF of X is $F(x)=\int_{-\infty}^{x} f(t)\,dt$, where $f(t)=\begin{cases}\frac{4}{9}t, & 1<t<2\\ \frac{1}{3}, & 2<t<3\\ 0, & \text{otherwise}\end{cases}$.

- If $x \leqslant 1, F(x)=\int_{-\infty}^{x} f(t)\,dt=0$
- If $1<x \leqslant 2, F(x)=\int_{-\infty}^{1} 0dt+\int_{1}^{x} \frac{4}{9}tdt=\frac{2}{9}(x^2-1)$
- If $2<x \leqslant 3, F(x)=\int_{-\infty}^{1} 0dt+\int_{1}^{2} \frac{4}{9}tdt+\int_{2}^{x} \frac{1}{3}dt=\frac{1}{3}x$
- If $x>3, F(x)=\int_{-\infty}^{1} 0dt+\int_{1}^{2} \frac{4}{9}tdt+\int_{2}^{3} \frac{1}{3}dt=1$

The CDF of X is $F(x)=\begin{cases}0, & x\leqslant 1\\ \frac{2}{9}(x^2-1), & 1<x\leqslant 2\\ \frac{1}{3}x, & 2<x\leqslant 3\\ 1, & x>3\end{cases}$.

Problem 2.34 The PF of X is

X	−2	0	2	3
P	0.2	0.2	0.5	0.1

Find the PF of $Y=X^2+1$.

 **Solution:**

The possible values of $Y=X^2+1$ are 1, 5, 10.

$P(Y=1)=P(X^2+1=1)=P(X=0)=0.2$.

$P(Y=5)=P(X^2+1=5)=P(X=2 \text{ or } X=-2)=P(X=2)+P(X=-2)=0.7$.

$P(Y=10)=P(X^2+1=10)=P(X=3 \text{ or } X=-3)=P(X=3)+P(X=-3)=0.1$.

The PF of Y is

Y	1	5	10
P	0.2	0.7	0.1

Problem 2.35 Suppose that X is exponentially distributed with the following PDF

$$f(x)=\begin{cases}\frac{1}{2}e^{-\frac{x}{2}}, & x>0\\ 0, & x\leqslant 0\end{cases}.$$

Find the PDF of (1) $Y=\frac{1}{X}$, (2) $Y=X^2$, (3) $Y=\sqrt{X}$.

Solution:

(1) $y=r(x)=\frac{1}{x}$ is one-to-one for $x>0$ and has its inverse function $s(y)=\frac{1}{y}, y>0$. $\frac{ds(y)}{dy}=-\frac{1}{y^2}$. The PDF of Y is

$$g(y)=\begin{cases}\frac{1}{2}e^{-\frac{1}{2y}}\times\left|-\frac{1}{y^2}\right|, & y>0\\ 0, & \text{otherwise}\end{cases}$$
$$=\begin{cases}\frac{1}{2y^2}e^{-\frac{1}{2y}}, & y>0\\ 0, & \text{otherwise}\end{cases}.$$

(2) $y=r(x)=x^2$ is one-to-one for $x>0$ and has its inverse function $s(y)=\sqrt{y}, y>0$. $\frac{ds(y)}{dy}=\frac{1}{2\sqrt{y}}$. The PDF of Y is

$$g(y)=\begin{cases}\frac{1}{2}e^{-\frac{\sqrt{y}}{2}}\times\left|\frac{1}{2\sqrt{y}}\right|, & y>0\\ 0, & \text{otherwise}\end{cases}$$
$$=\begin{cases}\frac{1}{4\sqrt{y}}e^{-\frac{\sqrt{y}}{2}}, & y>0\\ 0, & \text{otherwise}\end{cases}.$$

(3) $y=r(x)=\sqrt{x}$ is one-to-one for $x>0$ and has its inverse function $s(y)=y^2, y>0$. $\frac{ds(y)}{dy}=2y$. The PDF of Y is

$$g(y)=\begin{cases}\frac{1}{2}e^{-\frac{y^2}{2}}\times|2y|, & y>0\\ 0, & \text{otherwise}\end{cases}$$

$$=\begin{cases} ye^{-\frac{y^2}{2}}, & y>0 \\ 0, & \text{otherwise} \end{cases}.$$

Problem 2.36 Suppose that the PDF of X is

$$f(x)=\begin{cases} 2x^3e^{-x^2}, & x>0 \\ 0, & x\leqslant 0 \end{cases}.$$

Find the PDF of (1) $Y=2X+4$, (2) $Y=X^2$, (3) $Y=\ln X$.

Solution:

(1) $y=2x+4$ is strictly increasing, and its inverse function is $s(y)=\frac{y-4}{2}$, $\frac{ds(y)}{dy}=\frac{1}{2}$. The support of X is $(0,+\infty)$, then the range of Y is $(4,+\infty)$. The PDF of Y is

$$f_Y(y)=\begin{cases} 2\left(\frac{y-4}{2}\right)^3 e^{-\left(\frac{y-4}{2}\right)^2}\times\left|\frac{1}{2}\right|, & y>4 \\ 0, & y\leqslant 4 \end{cases}$$

$$=\begin{cases} \left(\frac{y-4}{2}\right)^3 e^{-\left(\frac{y-4}{2}\right)^2}, & y>4 \\ 0, & y\leqslant 4 \end{cases}.$$

(2) $y=x^2$ is strictly increasing in $(0,+\infty)$, and its inverse function is $s(y)=y^{\frac{1}{2}}$, $\frac{ds(y)}{dy}=\frac{1}{2\sqrt{y}}$. The support of X is $(0,+\infty)$, then the range of Y is $(0,+\infty)$. The PDF of Y is

$$f_Y(y)=\begin{cases} 2\left(y^{\frac{1}{2}}\right)^3 e^{-\left(y^{\frac{1}{2}}\right)^2}\times\left|\frac{1}{2\sqrt{y}}\right|, & y>0 \\ 0, & y\leqslant 0 \end{cases}$$

$$=\begin{cases} ye^{-y}, & y>0 \\ 0, & y\leqslant 0 \end{cases}.$$

(3) $y=\ln x$ is strictly increasing in $(0,+\infty)$, and its inverse function is $s(y)=e^y$, $\frac{ds(y)}{dy}=e^y$. The support of X is $(0,+\infty)$, then the range of Y is $(-\infty,+\infty)$. The PDF of Y is

$$f_Y(y)=2(e^y)^3e^{-(e^y)^2}\times|e^y|=2e^{4y}e^{-e^{2y}}=2e^{4y-e^{2y}}, -\infty<y<+\infty.$$

Problem 2.37 Suppose $X\sim U(-2,5)$, find the PDF of $Y=X^2$.

Solution 1:

The PDF of X is $f(x)=\begin{cases} \frac{1}{7}, & -2<x<5 \\ 0, & \text{otherwise} \end{cases}$.

The CDF of Y is $G(y)=P(Y\leqslant y)=P(X^2\leqslant y)$. Since $Y=X^2$, $-2<X<5$, then $0\leqslant Y<25$, as shown in Figure 2.9.

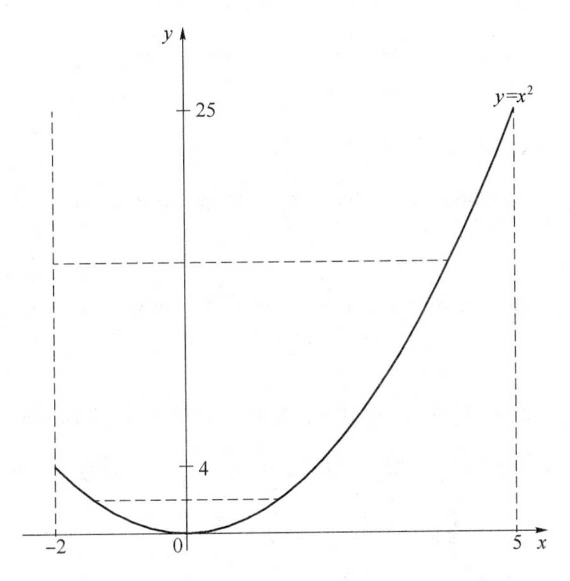

Figure 2.9

- If $y<0$, $G(y)=P(\varnothing)=0$
- If $0\leqslant y<4$,

$$
\begin{aligned}
G(y) &= P(X^2\leqslant y)\\
&= P(-\sqrt{y}\leqslant X\leqslant\sqrt{y})\\
&= \int_{-\sqrt{y}}^{\sqrt{y}}\frac{1}{7}\mathrm{d}x\\
&= \frac{2\sqrt{y}}{7}
\end{aligned}
$$

- If $4\leqslant y<25$,

$$
\begin{aligned}
G(y) &= P(X^2\leqslant y)\\
&= P(-2<X\leqslant\sqrt{y})\\
&= \int_{-2}^{\sqrt{y}}\frac{1}{7}\mathrm{d}x\\
&= \frac{\sqrt{y}+2}{7}
\end{aligned}
$$

- If $25\leqslant y$, $G(y)=P(\Omega)=1$

It follows that the PDF of Y is

$$
g(y)=G'(y)=\begin{cases}\dfrac{1}{7\sqrt{y}}, & 0\leqslant y<4\\[2mm] \dfrac{1}{14\sqrt{y}}, & 4\leqslant y<25\\[2mm] 0, & \text{otherwise}\end{cases}.
$$

Solution 2:

Note that the transformation $y=x^2$ is not strictly monotonic on $(-2,5)$. Therefore, we cannot apply the Change of Variable Theorem straight away. More precisely, $y=x^2$ is two-to-one on $(-2,2)$ and one-to-one on $[2,5)$.

- Let us focus on the interval $(-2,2)$. Noting that

(i) $y=x^2$ is strictly monotonic decreasing and differentiable on $(-2,0)$ and has its inverse function $s_1(y)=-\sqrt{y}$.

(ii) $y=x^2$ is strictly monotonic increasing and differentiable on $(0,2)$ and has its inverse function $s_2(y)=\sqrt{y}$.

If $-2<x<2$, then $y=x^2\in[0,4)$. By the general Change of Variable Theorem, for $0\leqslant y<4$,

$$\begin{aligned}g(y)&=f[s_1(y)]\times|s_1'(y)|+f[s_2(y)]\times|s_2'(y)|\\&=\frac{1}{7}\times\left|-\frac{1}{2\sqrt{y}}\right|+\frac{1}{7}\times\left|\frac{1}{2\sqrt{y}}\right|\\&=\frac{1}{7\sqrt{y}}.\end{aligned}$$

- On the interval $[2,5)$, $y=x^2$ is strictly monotonic increasing and has its inverse function $s(y)=\sqrt{y}$.

If $2\leqslant x<5$, then $y=x^2\in[4,25)$. Thus we can apply the Change of Variable Theorem.

For $4\leqslant y<25$,

$$\begin{aligned}g(y)&=f[s(y)]\times|s'(y)|\\&=\frac{1}{7}\times\left|\frac{1}{2\sqrt{y}}\right|\\&=\frac{1}{14\sqrt{y}}.\end{aligned}$$

It follows that the PDF of Y is

$$g(y)=G'(y)=\begin{cases}\dfrac{1}{7\sqrt{y}}, & 0\leqslant y<4\\ \dfrac{1}{14\sqrt{y}}, & 4\leqslant y<25\\ 0, & \text{otherwise}\end{cases}.$$

Problem 2.38 Suppose $\ln X\sim N(1,2^2)$, find $P\left(\frac{1}{2}<X<2\right)$.

Solution:

$$P\left(\frac{1}{2}<X<2\right)=P(-\ln 2<\ln X<\ln 2)$$
$$=P\left(\frac{-\ln 2-1}{2}<\frac{\ln X-1}{2}<\frac{\ln 2-1}{2}\right)$$
$$=\Phi\left(\frac{\ln 2-1}{2}\right)-\Phi\left(\frac{-\ln 2-1}{2}\right)$$
$$=\Phi(-0.15)-\Phi(-0.85)$$
$$=\Phi(0.85)-\Phi(0.15)$$
$$=0.802\ 3-0.559\ 6$$
$$=0.242\ 7.$$

Problem 2.39 Suppose that the PDF of a random variable X is as follows:

$$f(x)=\begin{cases}(c^2+c)x, & 1\leqslant x\leqslant\sqrt{\frac{3}{2}} \\ 0, & \text{otherwise}\end{cases}.$$

(1) Determine $c>0$ that renders $f(x)$ a valid density function.

(2) Find $P(X<1.1)$.

(3) Find $P(X=1.1)$

Solution:

(1)
$$\int_{-\infty}^{+\infty} f(x)\,\mathrm{d}x=\int_1^{\sqrt{\frac{3}{2}}}(c^2+c)x\,\mathrm{d}x=(c^2+c)\times\frac{1}{4}=1.$$

Therefore, $c^2+c=4$, then $c=\frac{\sqrt{17}-1}{2}$.

$$f(x)=\begin{cases}4x, & 1\leqslant x\leqslant\sqrt{\frac{3}{2}} \\ 0, & \text{otherwise}\end{cases}.$$

(2)
$$P(X<1.1)=\int_1^{1.1}4x\,\mathrm{d}x=2x^2\Big|_1^{1.1}=0.42.$$

(3) Since X is a continuous random variable, $P(X=1.1)=0$.

Problem 2.40 Suppose that X is exponentially distributed with the following PDF

$$f(x)=\begin{cases}\lambda\mathrm{e}^{-\lambda x}, & x>0 \\ 0, & x\leqslant 0\end{cases},$$

where $\lambda>0$. Let $Y=\min\{2,X\}$. Find the CDF of Y.

Solution:

$$F_Y(y)=P(Y\leqslant y)=P(\min\{2,X\}\leqslant y)$$

- If $y\leqslant 0$, $F_Y(y)=0$.
- If $0<y<2$,

$$\begin{aligned}F_Y(y)&=P(\min\{2,X\}\leqslant y)\\&=P(X\leqslant y)\\&=\int_0^y \lambda\,\mathrm{e}^{-\lambda x}\mathrm{d}x\\&=1-\mathrm{e}^{-\lambda y}.\end{aligned}$$

- If $2\leqslant y$, $F_Y(y)=1$.

Therefore, $F_Y(y)=\begin{cases}0, & y\leqslant 0\\ 1-\mathrm{e}^{-\lambda y}, & 0<y<2\\ 1, & y\geqslant 2\end{cases}$.

Note: the random variable Y in this problem is neither continuous nor discrete.

Chapter 3 Multivariate Probability Distributions

Summary of Knowledge

Exercise Solutions

Problem 3.1 Suppose that 3 balls are chosen without replacement from an urn consisting of 5 white and 8 red balls. Let X_i equal 1 if the i-th ball selected is white and let it equal 0 otherwise. Find the joint PF of (X_1, X_2).

Solution:

$$P(X_1=0, X_2=0)=\frac{8}{13}\times\frac{7}{12}=\frac{14}{39}$$

$$P(X_1=0, X_2=1)=\frac{8}{13}\times\frac{5}{12}=\frac{10}{39}$$

$$P(X_1=1, X_2=0)=\frac{5}{13}\times\frac{8}{12}=\frac{10}{39}$$

$$P(X_1=1, X_2=1)=\frac{5}{13}\times\frac{4}{12}=\frac{5}{39}$$

Therefore, the joint PF of (X_1, X_2) is given in the following table.

	$X_2=0$	$X_2=1$
$X_1=0$	$\frac{14}{39}$	$\frac{10}{39}$
$X_1=1$	$\frac{10}{39}$	$\frac{5}{39}$

Problem 3.2 Two balanced dice are rolled. Let X be the larger of the two values shown on the dice, and let Y be the absolute value of the difference of the two values shown. Give the joint PF of X and Y.

Solution:

The sample space contains the following 36 outcomes.

(1,1),(1,2),(1,3),(1,4),(1,5),(1,6)

(2,1),(2,2),(2,3),(2,4),(2,5),(2,6)

(3,1),(3,2),(3,3),(3,4),(3,5),(3,6)

(4,1),(4,2),(4,3),(4,4),(4,5),(4,6)

(5,1),(5,2),(5,3),(5,4),(5,5),(5,6)

(6,1),(6,2),(6,3),(6,4),(6,5),(6,6)

where (m,n) is the outcome in which the first die shows face m and the second die shows face n.

X is the larger of the two values shown on the dice and takes 1,2,3,4, 5,6. Y is the absolute value of the difference of the two values shown and takes 0,1,2,3,4,5.

$P(X=1,Y=0)=P(\{(1,1)\})=\frac{1}{36}$. Similarly, $P(X=i,Y=0)=P(\{(i,i)\})=\frac{1}{36}$, for $i=1,2,\cdots,6$.

$P(X=1,Y=1)=P(X=1,Y=2)=\cdots=P(X=1,Y=5)=0$, because if the larger of the two values is 1, the difference can not be 1 or more.

$P(X=2,Y=1)=P(\{(2,1),(1,2)\})=\frac{2}{36}=\frac{1}{18}$. Similarly, $P(X=3,Y=1)=P(X=3,Y=2)=P(X=4,Y=1)=P(X=4,Y=2)=P(X=4,Y=3)=P(X=5,Y=1)=\cdots=P(X=5,Y=4)=P(X=6,Y=1)=\cdots=P(X=6,Y=5)=\frac{2}{36}=\frac{1}{18}$, since each corresponding event is a subset of two outcomes from the sample space.

All other values of $P(X=i,Y=j)$ are 0.

Therefore, the joint PF of (X,Y) is given in the following table.

	$Y=0$	$Y=1$	$Y=2$	$Y=3$	$Y=4$	$Y=5$
$X=1$	$\frac{1}{36}$	0	0	0	0	0
$X=2$	$\frac{1}{36}$	$\frac{1}{18}$	0	0	0	0
$X=3$	$\frac{1}{36}$	$\frac{1}{18}$	$\frac{1}{18}$	0	0	0
$X=4$	$\frac{1}{36}$	$\frac{1}{18}$	$\frac{1}{18}$	$\frac{1}{18}$	0	0
$X=5$	$\frac{1}{36}$	$\frac{1}{18}$	$\frac{1}{18}$	$\frac{1}{18}$	$\frac{1}{18}$	0
$X=6$	$\frac{1}{36}$	$\frac{1}{18}$	$\frac{1}{18}$	$\frac{1}{18}$	$\frac{1}{18}$	$\frac{1}{18}$

Problem 3.3 Toss an unbiased coin four times. Let X be the number of heads in the first three tosses and Y be the number of heads in the last two tosses. Find the joint PF of (X,Y).

Solution:

When a fair coin is tossed four times, then the total number of outcomes (an equally likely outcome) will be $2^4=16$. X is the random variable for the number of heads in the first three tosses and Y is the random variable for the number of heads in the last two tosses. X takes $0,1,2$, or 3, and Y takes $0,1$, or 2.

$P(X=0,Y=0)=P(\{\text{TTTT}\})=\frac{1}{16}$,

$P(X=0,Y=1)=P(\{\text{TTTH}\})=\frac{1}{16}$,

$P(X=0,Y=2)=P(\varnothing)=0$,

$P(X=1,Y=0)=P(\{\text{HTTT}\},\{\text{THTT}\})=\frac{2}{16}=\frac{1}{8}$,

$P(X=1,Y=1)=P(\{\text{HTTH}\},\{\text{THTH}\},\{\text{TTHT}\})=\frac{3}{16}$,

$P(X=1,Y=2)=P(\{\text{TTHH}\})=\frac{1}{16}$,

$P(X=2,Y=0)=P(\{\text{HHTT}\})=\frac{1}{16}$,

$P(X=2,Y=1)=P(\{\text{HHTH}\},\{\text{HTHT}\},\{\text{THHT}\})=\frac{3}{16}$,

$P(X=2,Y=2)=P(\{\text{HTHH}\},\{\text{THHH}\})=\frac{2}{16}=\frac{1}{8}$,

$P(X=3,Y=0)=P(\varnothing)=0$,

$P(X=3,Y=1)=P(\{\text{HHHT}\})=\frac{1}{16}$,

$P(X=3,Y=2)=P(\{\text{HHHH}\})=\frac{1}{16}$.

Therefore, the joint PF of (X,Y) is given in the following table.

	$Y=0$	$Y=1$	$Y=2$
$X=0$	$\frac{1}{16}$	$\frac{1}{16}$	0
$X=1$	$\frac{1}{8}$	$\frac{3}{16}$	$\frac{1}{16}$
$X=2$	$\frac{1}{16}$	$\frac{3}{16}$	$\frac{1}{8}$
$X=3$	0	$\frac{1}{16}$	$\frac{1}{16}$

Problem 3.4 A bag contains 5 chocolate-chip muffins, 3 blueberry muffins and 2 lemon muffins. Choose three muffins from this bag at random. Let X be the number of lemon muffins chosen and let Y be the number of blueberry muffins chosen. Find the joint PF of (X,Y).

 **Solution**:

X takes 0,1, or 2. Y takes 0,1,2, or 3.

$$P(X=0,Y=0)=\frac{C_5^3}{C_{10}^3}=\frac{10}{120}=\frac{1}{12},$$

$$P(X=0,Y=1)=\frac{C_5^2C_3^1}{C_{10}^3}=\frac{30}{120}=\frac{1}{4},$$

$$P(X=0,Y=2)=\frac{C_5^1C_3^2}{C_{10}^3}=\frac{15}{120}=\frac{1}{8},$$

$$P(X=0,Y=3)=\frac{1}{C_{10}^3}=\frac{1}{120},$$

$$P(X=1,Y=0)=\frac{C_2^1C_5^2}{C_{10}^3}=\frac{20}{120}=\frac{1}{6},$$

$$P(X=1,Y=1)=\frac{C_2^1C_5^1C_3^1}{C_{10}^3}=\frac{30}{120}=\frac{1}{4},$$

$$P(X=1,Y=2)=\frac{C_2^1C_3^2}{C_{10}^3}=\frac{6}{120}=\frac{1}{20},$$

$$P(X=2,Y=0)=\frac{C_5^1}{C_{10}^3}=\frac{5}{120}=\frac{1}{24},$$

$$P(X=2,Y=1)=\frac{C_3^1}{C_{10}^3}=\frac{3}{120}=\frac{1}{40},$$

$P(X=1,Y=3)=P(X=2,Y=2)=P(X=2,Y=3)=0.$

Hence, the joint PF of (X,Y) is as follows:

	$Y=0$	$Y=1$	$Y=2$	$Y=3$
$X=0$	$\frac{1}{12}$	$\frac{1}{4}$	$\frac{1}{8}$	$\frac{1}{120}$
$X=1$	$\frac{1}{6}$	$\frac{1}{4}$	$\frac{1}{20}$	0
$X=2$	$\frac{1}{24}$	$\frac{1}{40}$	0	0

Problem 3.5 Suppose that 2 batteries are randomly chosen without replacement from the following group of 12 batteries: 3 new, 4 used, 5 defective. Let X denote the number of new batteries chosen. Let Y denote the number of used batteries chosen. Find the joint PF of (X,Y).

Solution:

X takes 0, 1, or 2. Y takes 0, 1, or 2.

$$P(X=0,Y=0)=\frac{C_5^2}{C_{12}^2}=\frac{10}{66}=\frac{5}{33},$$

$$P(X=0,Y=1)=\frac{C_4^1C_5^1}{C_{12}^2}=\frac{20}{66}=\frac{10}{33},$$

$$P(X=0,Y=2)=\frac{C_4^2}{C_{12}^2}=\frac{6}{66}=\frac{1}{11},$$

$$P(X=1,Y=0)=\frac{C_3^1C_5^1}{C_{12}^2}=\frac{15}{66}=\frac{5}{22},$$

$$P(X=1,Y=1)=\frac{C_3^1C_4^1}{C_{12}^2}=\frac{12}{66}=\frac{2}{11},$$

$$P(X=2,Y=0)=\frac{C_3^2}{C_{12}^2}=\frac{3}{66}=\frac{1}{22},$$

$P(X=1,Y=2)=P(X=2,Y=1)=P(X=2,Y=2)=0.$

Hence, the joint PF of (X,Y) is as follows:

	$Y=0$	$Y=1$	$Y=2$
$X=0$	$\frac{5}{33}$	$\frac{10}{33}$	$\frac{1}{11}$
$X=1$	$\frac{5}{22}$	$\frac{2}{11}$	0
$X=2$	$\frac{1}{22}$	0	0

Problem 3.6 Suppose the joint PDF of two random variables X and Y is given by

$$f(x,y)=\begin{cases} C\mathrm{e}^{-(x+y)}, & x>0,y>0 \\ 0, & \text{otherwise} \end{cases}.$$

Find:

(1) the constant C;

(2) the joint CDF $F(x,y)$;

(3) $P(X>1)$;

(4) $P(X\geqslant Y)$;

(5) $P(X+2Y\leqslant 1)$.

Solution:

(1) The support of (X,Y) is $\{(x,y)\mid x>0,y>0\}$, as shown in Figure 3.1.

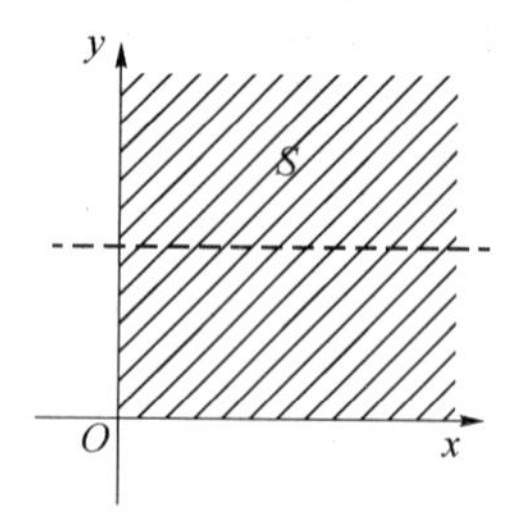

Figure 3.1

$$\begin{aligned}1&=\int_{-\infty}^{+\infty}\int_{-\infty}^{+\infty}f(x,y)\,\mathrm{d}x\mathrm{d}y\\&=\int_{0}^{+\infty}\int_{0}^{+\infty}C\,\mathrm{e}^{-(x+y)}\,\mathrm{d}x\mathrm{d}y\\&=\int_{0}^{+\infty}\left[\int_{0}^{+\infty}C\,\mathrm{e}^{-(x+y)}\,\mathrm{d}x\right]\mathrm{d}y\\&=\int_{0}^{+\infty}C\,\mathrm{e}^{-y}\,\mathrm{d}y\\&=C.\end{aligned}$$

Therefore, $C=1$. The joint PDF of (X,Y) is

$$f(x,y)=\begin{cases}\mathrm{e}^{-(x+y)}, & x>0,y>0\\0, & \text{otherwise}\end{cases}.$$

(2) See Figure 3.2.

- If $x\leqslant 0$ or $y\leqslant 0$, $F(x,y)=0$.

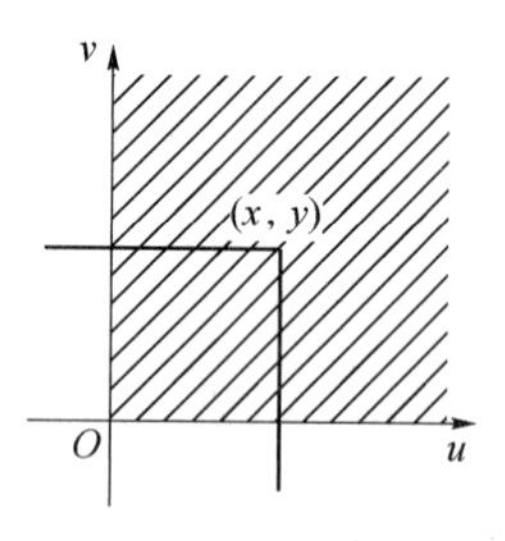

Figure 3.2

- If $x>0,y>0$

$$\begin{aligned}F(x,y)&=P(X\leqslant x,Y\leqslant y)\\&=\int_{-\infty}^{x}\int_{-\infty}^{y}f(u,v)\,\mathrm{d}u\mathrm{d}v\\&=\int_{0}^{x}\int_{0}^{y}\mathrm{e}^{-(u+v)}\,\mathrm{d}u\mathrm{d}v\end{aligned}$$

$$= \int_0^y \left[\int_0^x e^{-(u+v)} du \right] dv$$
$$= (1-e^{-x})(1-e^{-y}).$$

Therefore, the joint CDF is

$$F(x,y) = \begin{cases} (1-e^{-x})(1-e^{-y}), & x>0, y>0 \\ 0, & \text{otherwise} \end{cases}.$$

(3)

$$\begin{aligned} P(X>1) &= 1-P(X \leqslant 1) \\ &= 1-P(X \leqslant 1, Y<+\infty) \\ &= 1-F(1,+\infty) \\ &= e^{-1}. \end{aligned}$$

(4) See Figure 3.3.

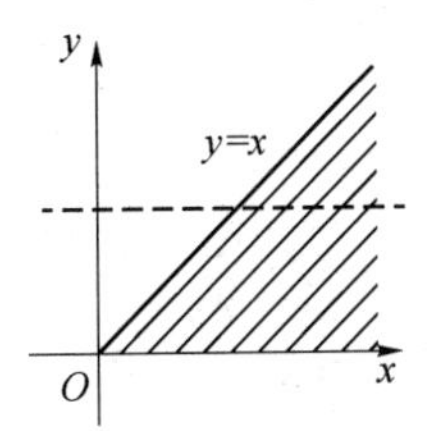

Figure 3.3

$$\begin{aligned} P(X \geqslant Y) &= \iint_{x \geqslant y} f(x,y) dxdy \\ &= \int_0^{+\infty} \left[\int_y^{+\infty} e^{-(x+y)} dx \right] dy \\ &= \int_0^{+\infty} e^{-2y} dy \\ &= \frac{1}{2}. \end{aligned}$$

(5) See Figure 3.4.

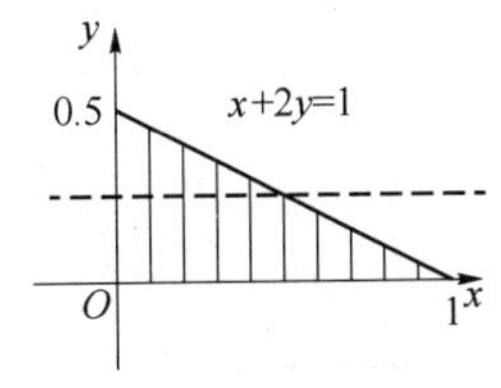

Figure 3.4

$$\begin{aligned} P(X+2Y \leqslant 1) &= \iint_{x+2y \leqslant 1} f(x,y) dxdy \\ &= \int_0^{0.5} \left[\int_0^{1-2y} e^{-(x+y)} dx \right] dy \end{aligned}$$

$$
\begin{aligned}
&= \int_0^{0.5} e^{-y}(1 - e^{2y-1})\,dy \\
&= -e^{-y} - e^{y-1} \Big|_0^{0.5} \\
&= 1 + e^{-1} - 2e^{-\frac{1}{2}}.
\end{aligned}
$$

Problem 3.7 Let the joint PDF of two random variable X and Y be given by

$$
f(x,y)=\begin{cases} C\,e^{-2(x+y)}, & x>0, y>0 \\ 0, & \text{otherwise} \end{cases}.
$$

Find:

(1) the constant C;

(2) the joint CDF $F(x,y)$;

(3) the marginal CDFs and PDFs of X and Y;

(4) the probability that (X,Y) falls into the region $\{(x,y) \mid x+y<1\}$.

Solution:

(1) The support of (X,Y) is $\{(x,y) \mid x>0, y>0\}$, as shown in Figure 3.5.

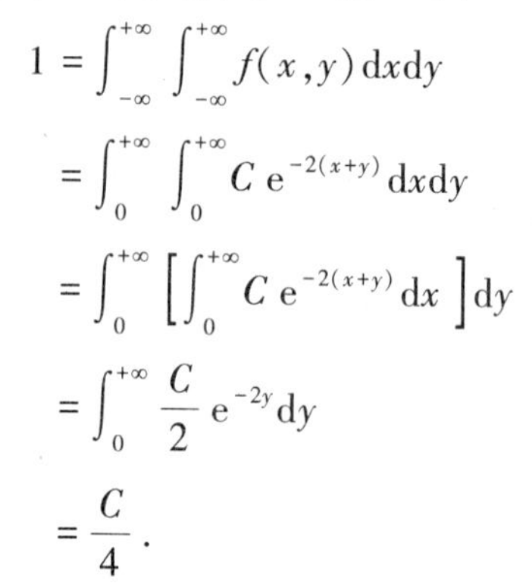

$$
\begin{aligned}
1 &= \int_{-\infty}^{+\infty}\int_{-\infty}^{+\infty} f(x,y)\,dx\,dy \\
&= \int_0^{+\infty}\int_0^{+\infty} C\,e^{-2(x+y)}\,dx\,dy \\
&= \int_0^{+\infty}\left[\int_0^{+\infty} C\,e^{-2(x+y)}\,dx\right]dy \\
&= \int_0^{+\infty} \frac{C}{2} e^{-2y}\,dy \\
&= \frac{C}{4}.
\end{aligned}
$$

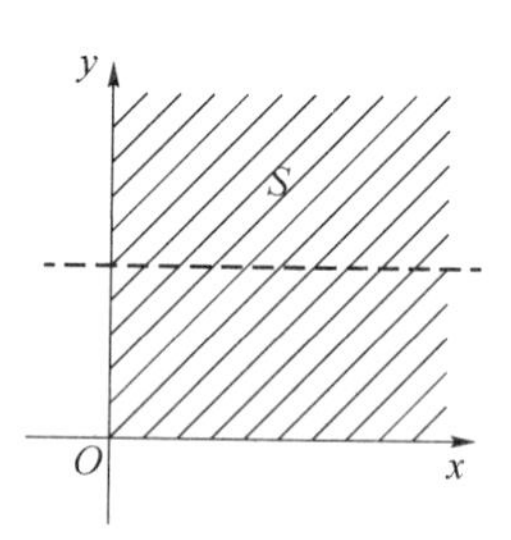

Figure 3.5

Therefore, $C=4$. The joint PDF of (X,Y) is

$$
f(x,y)=\begin{cases} 4\,e^{-2(x+y)}, & x>0, y>0 \\ 0, & \text{otherwise} \end{cases}.
$$

(2) See Figure 3.6.

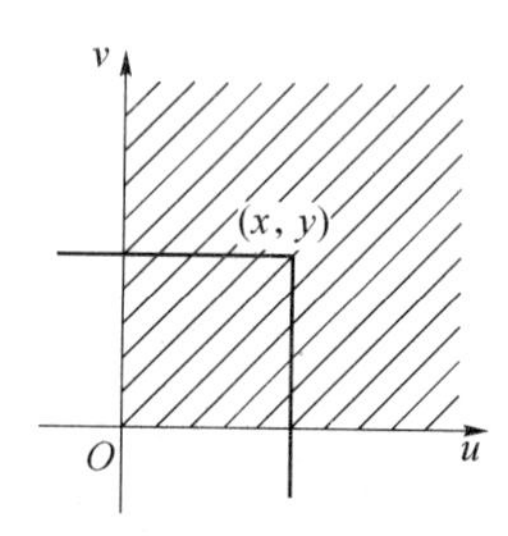

Figure 3.6

- If $x \leqslant 0$ or $y \leqslant 0, F(x,y) = 0$.
- If $x>0, y>0$,

$$\begin{aligned} F(x,y) &= P(X \leqslant x, Y \leqslant y) \\ &= \int_{-\infty}^{x} \int_{-\infty}^{y} f(u,v)\,\mathrm{d}u\mathrm{d}v \\ &= \int_{0}^{x} \int_{0}^{y} 4\,\mathrm{e}^{-2(u+v)}\,\mathrm{d}u\mathrm{d}v \\ &= \int_{0}^{y} \left[\int_{0}^{x} 4\,\mathrm{e}^{-2(u+v)}\,\mathrm{d}u \right] \mathrm{d}v \\ &= (1-\mathrm{e}^{-2x})(1-\mathrm{e}^{-2y}). \end{aligned}$$

The joint CDF of (X,Y) is

$$F(x,y) = \begin{cases} (1-\mathrm{e}^{-2x})(1-\mathrm{e}^{-2y}), & x>0, y>0 \\ 0, & \text{otherwise} \end{cases}.$$

(3)
$$F_X(x) = F(x,+\infty) = \begin{cases} 1-\mathrm{e}^{-2x}, & x>0 \\ 0, & \text{otherwise} \end{cases}.$$

$$f_X(x) = F_X'(x) = \begin{cases} 2\,\mathrm{e}^{-2x}, & x>0 \\ 0, & \text{otherwise} \end{cases}.$$

$$F_Y(y) = F(+\infty,y) = \begin{cases} 1-\mathrm{e}^{-2y}, & y>0 \\ 0, & \text{otherwise} \end{cases}.$$

$$f_Y(y) = F_Y'(y) = \begin{cases} 2\,\mathrm{e}^{-2y}, & y>0 \\ 0, & \text{otherwise} \end{cases}.$$

(4) See Figure 3.7.

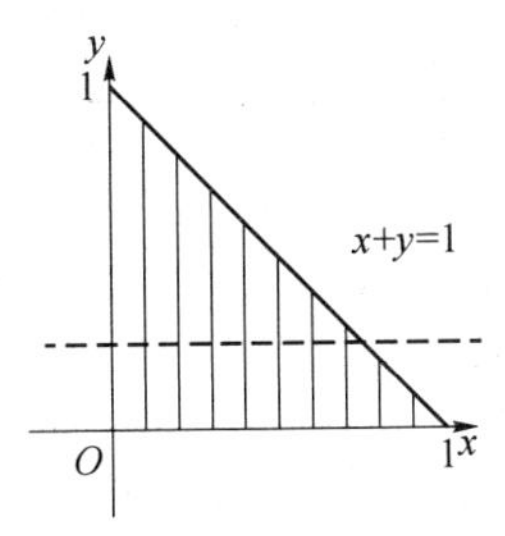

Figure 3.7

$$\begin{aligned} P(X+Y<1) &= \iint_{x+y<1} f(x,y)\,\mathrm{d}x\mathrm{d}y \\ &= \int_{0}^{1} \left[\int_{0}^{1-y} 4\,\mathrm{e}^{-2(x+y)}\,\mathrm{d}x \right] \mathrm{d}y \\ &= \int_{0}^{1} 2\,\mathrm{e}^{-2y}(1-\mathrm{e}^{2y-2})\,\mathrm{d}y \\ &= -\mathrm{e}^{-2y} - 2\,\mathrm{e}^{-2}y \Big|_{0}^{1} \\ &= 1-3\mathrm{e}^{-2}. \end{aligned}$$

Problem 3.8 Suppose the joint PDF of two random variable X and Y is

$$f(x,y)=\begin{cases} a\,\mathrm{e}^{-y}, & 0<x<y \\ 0, & \text{otherwise} \end{cases}.$$

Find:

(1) the constant a;

(2) the marginal PDFs of X and Y;

(3) $P(X+Y\leqslant 1)$.

Solution:

(1) The support S of (X,Y) is $\{(x,y)\mid 0<x<y\}$, as shown in Figure 3.8.

$$\begin{aligned} 1 &= \int_{-\infty}^{+\infty}\int_{-\infty}^{+\infty} f(x,y)\,\mathrm{d}x\mathrm{d}y \\ &= \iint_S a\,\mathrm{e}^{-y}\,\mathrm{d}x\mathrm{d}y \\ &= \int_0^{+\infty}\left(\int_0^y a\,\mathrm{e}^{-y}\,\mathrm{d}x\right)\mathrm{d}y \\ &= \int_0^{+\infty} ay\mathrm{e}^{-y}\,\mathrm{d}y \\ &= a. \end{aligned}$$

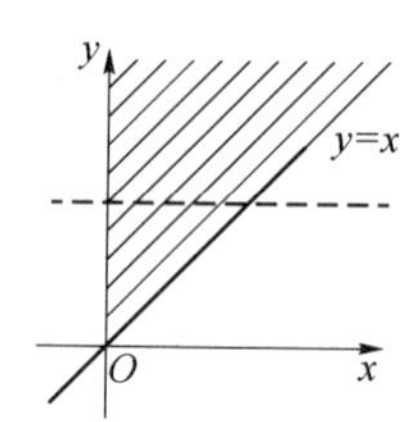

Figure 3.8

Therefore, $a=1$. The joint PDF is

$$f(x,y)=\begin{cases} \mathrm{e}^{-y}, & 0<x<y \\ 0, & \text{otherwise} \end{cases}.$$

(2) Find the marginal PDF of X, as shown in Figure 3.9.

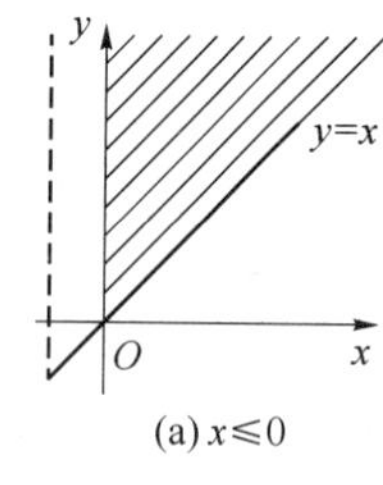

(a) $x\leqslant 0$

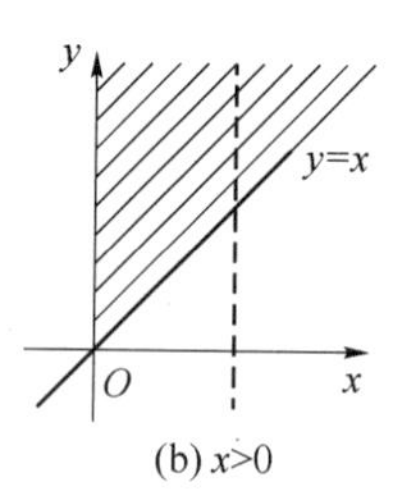

(b) $x>0$

Figure 3.9

- If $x\leqslant 0$, $f_X(x)=0$.
- If $x>0$,

$$\begin{aligned} f_X(x) &= \int_{-\infty}^{+\infty} f(x,y)\,\mathrm{d}y \\ &= \int_x^{+\infty} \mathrm{e}^{-y}\,\mathrm{d}y \\ &= \mathrm{e}^{-x}. \end{aligned}$$

The marginal PDF of X is

$$f_X(x)=\begin{cases}e^{-x}, & x>0\\ 0, & \text{otherwise}\end{cases}.$$

Find the marginal PDF of Y, as shown in Figure 3.10.

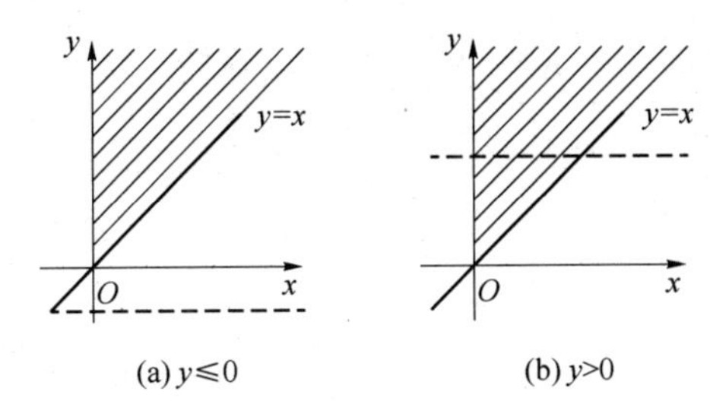

Figure 3.10

- If $y\leqslant 0, f_Y(y)=0$.
- If $y>0$,

$$\begin{aligned}f_Y(y)&=\int_{-\infty}^{+\infty}f(x,y)\,\mathrm{d}x\\ &=\int_0^y e^{-y}\mathrm{d}x\\ &=ye^{-y}.\end{aligned}$$

The marginal PDF of Y is

$$f_Y(y)=\begin{cases}ye^{-y}, & y>0\\ 0, & \text{otherwise}\end{cases}.$$

(3) See Figure 3.11.

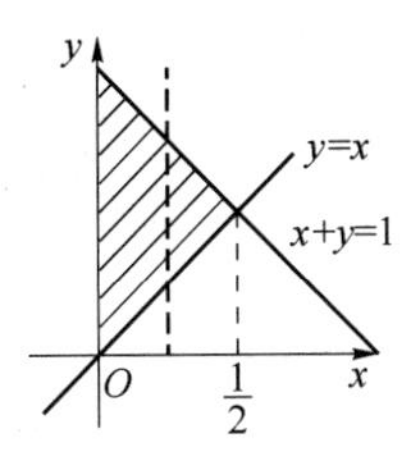

Figure 3.11

$$\begin{aligned}P(X+Y\leqslant 1)&=\iint_{x+y\leqslant 1}f(x,y)\,\mathrm{d}x\mathrm{d}y\\ &=\int_0^{1/2}\left(\int_x^{1-x}e^{-y}\mathrm{d}y\right)\mathrm{d}x\\ &=\int_0^{1/2}(e^{-x}-e^{x-1})\,\mathrm{d}x\\ &=-e^{-x}-e^{x-1}\Big|_0^{1/2}\\ &=1-2e^{-1/2}+e^{-1}.\end{aligned}$$

Problem 3.9 Suppose that (X,Y) follows the uniform distribution over the region $D=\{(x,y)\mid 0\leqslant y\leqslant 1-x^2\}$.

(1) What is the joint PDF of (X,Y)?

(2) What are the marginal PDFs of X and Y?

(3) What is the conditional PDF of Y when $X=-0.5$?

(4) Find $P[(X,Y)\in B]$, where $B=\{(x,y)\mid y\geqslant x^2\}$.

Solution:

(1) As shown in Figure 3.12, the area of D is $S_D=\int_{-1}^{1}(1-x^2)\mathrm{d}x=\frac{4}{3}$. The joint PDF of (X,Y) is

$$f(x,y)=\begin{cases}\frac{3}{4}, & (x,y)\in D\\ 0, & \text{otherwise}\end{cases}.$$

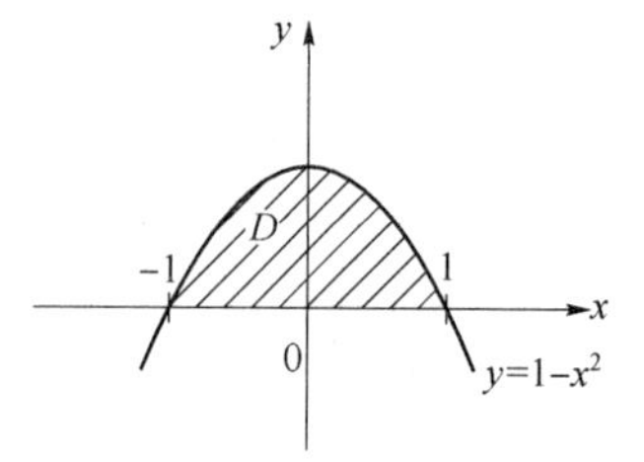

Figure 3.12

(2) Find the marginal PDF of X, as shown in Figure 3.13.

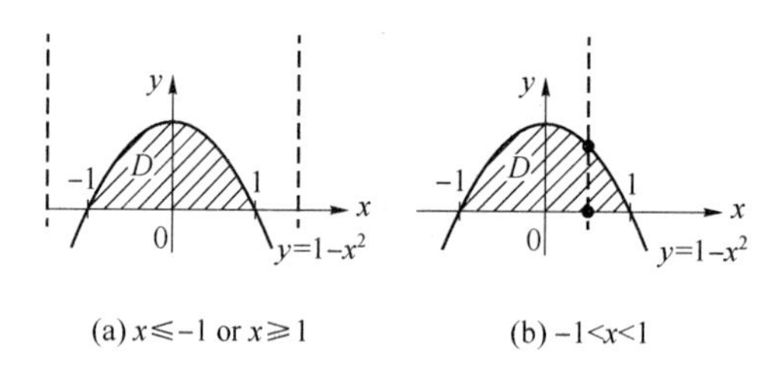

(a) $x\leqslant -1$ or $x\geqslant 1$ (b) $-1<x<1$

Figure 3.13

- If $x\leqslant -1$ or $x\geqslant 1$, $f_X(x)=0$.
- If $-1<x<1$,

$$\begin{aligned}f_X(x)&=\int_{-\infty}^{+\infty}f(x,y)\mathrm{d}y\\&=\int_{0}^{1-x^2}\frac{3}{4}\mathrm{d}y\\&=\frac{3}{4}(1-x^2).\end{aligned}$$

The marginal PDF of X is

$$f_X(x)=\begin{cases}\frac{3}{4}(1-x^2), & -1<x<1\\ 0, & \text{otherwise}\end{cases}.$$

Find the marginal PDF of Y, as shown in Figure 3.14.

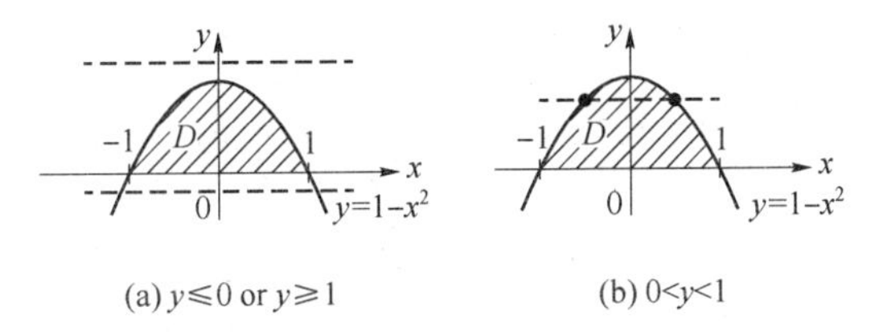

Figure 3.14

- If $y\leqslant 0$ or $y\geqslant 1, f_Y(y)=0$.
- If $0<y<1$,

$$\begin{aligned}f_Y(y)&=\int_{-\infty}^{+\infty}f(x,y)\,\mathrm{d}x\\&=\int_{-\sqrt{1-y}}^{\sqrt{1-y}}\frac{3}{4}\mathrm{d}x\\&=\frac{3\sqrt{1-y}}{2}.\end{aligned}$$

Therefore, the marginal PDF of Y is

$$f_Y(y)=\begin{cases}\frac{3\sqrt{1-y}}{2}, & 0<y<1\\ 0, & \text{otherwise}\end{cases}.$$

(3) The conditional PDF of Y when $X=-0.5$ is

$$\begin{aligned}f_{Y\mid X=-0.5}(y\mid x=-0.5)&=\frac{f(x=-0.5,y)}{f_X(x=-0.5)}\\&=\begin{cases}\dfrac{\frac{3}{4}}{\frac{3}{4}\times[1-(-0.5)^2]}, & 0\leqslant y\leqslant\frac{3}{4}\\ 0, & \text{otherwise}\end{cases}\\&=\begin{cases}\frac{4}{3}, & 0\leqslant y\leqslant\frac{3}{4}\\ 0, & \text{otherwise}\end{cases}.\end{aligned}$$

(4) See Figure 3.15.

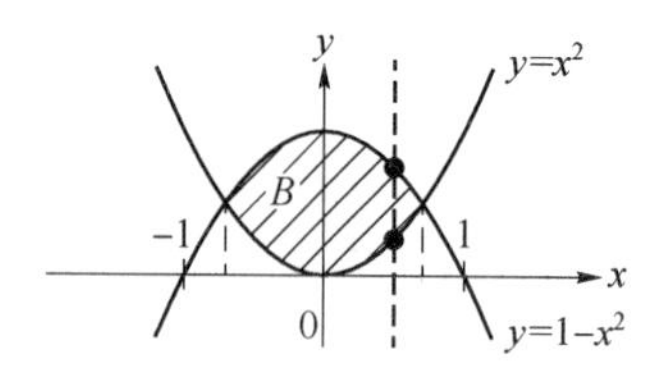

Figure 3.15

$$P[(X,Y)\in B]=\iint_B f(x,y)\,dxdy$$
$$=\int_{-\frac{\sqrt{2}}{2}}^{\frac{\sqrt{2}}{2}}\left(\int_{x^2}^{1-x^2}\frac{3}{4}dy\right)dx$$
$$=\frac{3}{4}\left(x-\frac{2x^3}{3}\right)\Bigg|_{-\frac{\sqrt{2}}{2}}^{\frac{\sqrt{2}}{2}}$$
$$=\frac{\sqrt{2}}{2}.$$

Problem 3.10 The joint PF of (X,Y) is

$$P(X=n,Y=m)=\frac{e^{-14}\times 7.14^m\times 6.86^{n-m}}{m!(n-m)!},m=0,1,\cdots,n;n=0,1,2,\cdots.$$

Find:

(1) the marginal PFs;

(2) the conditional PFs.

 **Solution**:

(1) The marginal PF of X is

$$P(X=n)=\sum_{m=0}^{n}P(X=n,Y=m)$$
$$=\sum_{m=0}^{n}\frac{e^{-14}\times 7.14^m\times 6.86^{n-m}}{m!(n-m)!}$$
$$=\frac{e^{-14}}{n!}\sum_{m=0}^{n}\frac{n!\times 7.14^m\times 6.86^{n-m}}{m!(n-m)!}$$
$$=\frac{e^{-14}}{n!}(7.14+6.86)^n$$
$$=\frac{e^{-14}\,14^n}{n!},n=0,1,2,\cdots.$$

X follows the Poisson distribution P(14).

The marginal PF of Y is

$$
\begin{aligned}
P(Y=m) &= \sum_{n=0}^{+\infty} P(X=n, Y=m) \\
&= \sum_{n=m}^{+\infty} \frac{e^{-14} \times 7.14^m \times 6.86^{n-m}}{m!(n-m)!} \\
&= \frac{e^{-14} \times 7.14^m}{m!} \sum_{n=m}^{+\infty} \frac{6.86^{n-m}}{(n-m)!} \qquad (\text{let } t = n-m) \\
&= \frac{e^{-14} \times 7.14^m}{m!} \sum_{t=0}^{+\infty} \frac{6.86^t}{t!} \\
&= \frac{e^{-14} \times 7.14^m}{m!} \times e^{6.86} \\
&= \frac{7.14^m}{m!} e^{-7.14}, m = 0,1,2,\cdots.
\end{aligned}
$$

Y follows P(7.14).

(2) The conditional PF of X when $Y=m$ is

$$
\begin{aligned}
P(X=n \mid Y=m) &= \frac{P(X=n, Y=m)}{P(Y=m)} \\
&= \frac{\dfrac{e^{-14} \times 7.14^m \times 6.86^{n-m}}{m!(n-m)!}}{\dfrac{7.14^m}{m!} e^{-7.14}} \\
&= \frac{6.86^{n-m}}{(n-m)!} e^{-6.86}, n=m, m+1, \cdots.
\end{aligned}
$$

The conditional PF of Y when $X=n$ is

$$
\begin{aligned}
P(Y=m \mid X=n) &= \frac{P(X=n, Y=m)}{P(X=n)} \\
&= \frac{\dfrac{e^{-14} \times 7.14^m \times 6.86^{n-m}}{m!(n-m)!}}{\dfrac{e^{-14} 14^n}{n!}} \\
&= C_n^m \left(\frac{7.14}{14}\right)^m \left(\frac{6.86}{14}\right)^{n-m}, m=0,1,\cdots,n.
\end{aligned}
$$

Problem 3.11 Let the joint PDF of two random variable X and Y be given by

$$
f(x,y) = \begin{cases} 4xy, & 0 \leqslant x < 1, 0 \leqslant y < 1 \\ 0, & \text{otherwise} \end{cases}.
$$

Find:

(1) the conditional PDFs $f_{X|Y}(x|y)$ and $f_{Y|X}(y|x)$;

(2) the joint CDF $F(x,y)$.

Solution:

(1) Find the marginal PDF of X, as shown in Figure 3.16.

- If $x \leqslant 0$ or $x \geqslant 1$, $f_X(x) = 0$.
- If $0<x<1$, $f_X(x) = \int_{-\infty}^{+\infty} f(x,y)\,\mathrm{d}y = \int_0^1 4xy\,\mathrm{d}y = 2x$.

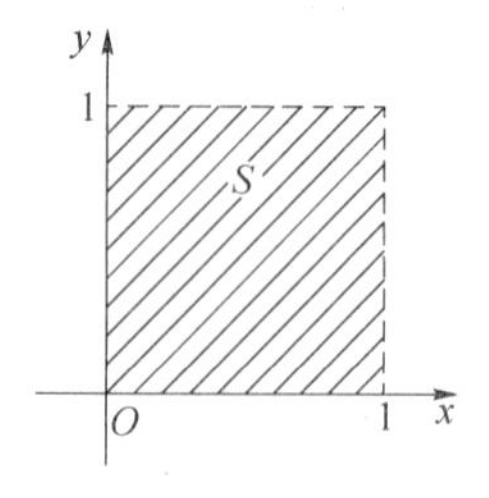

Figure 3.16

Therefore, the marginal PDF of X is

$$f_X(x) = \begin{cases} 2x, & 0<x<1 \\ 0, & \text{otherwise} \end{cases}.$$

By the symmetry, the marginal PDF of Y is

$$f_Y(y) = \begin{cases} 2y, & 0<y<1 \\ 0, & \text{otherwise} \end{cases}.$$

Therefore,

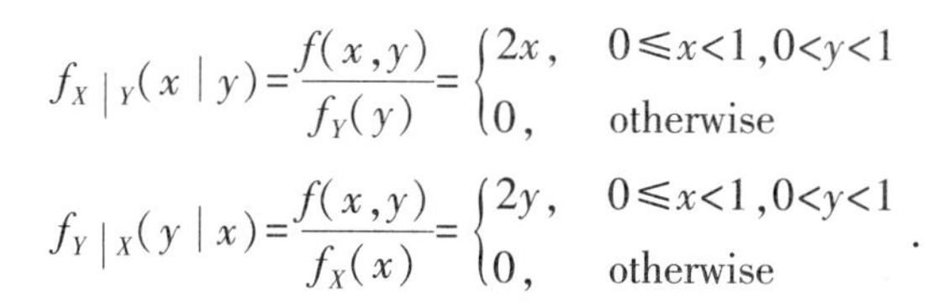

$$f_{X|Y}(x|y) = \frac{f(x,y)}{f_Y(y)} = \begin{cases} 2x, & 0 \leqslant x<1, 0<y<1 \\ 0, & \text{otherwise} \end{cases}$$

$$f_{Y|X}(y|x) = \frac{f(x,y)}{f_X(x)} = \begin{cases} 2y, & 0 \leqslant x<1, 0<y<1 \\ 0, & \text{otherwise} \end{cases}.$$

(2) $F(x,y) = \int_{-\infty}^{x}\int_{-\infty}^{y} f(u,v)\,\mathrm{d}u\mathrm{d}v$, where

$$f(u,v) = \begin{cases} 4uv, & 0 \leqslant u<1, 0 \leqslant v<1 \\ 0, & \text{otherwise} \end{cases}.$$

- If $x \leqslant 0$ or $y \leqslant 0$, $F(x,y) = 0$.
- If $0<x \leqslant 1, 0<y \leqslant 1$, $F(x,y) = \int_0^x\int_0^y 4uv\,\mathrm{d}u\mathrm{d}v = x^2 y^2$.
- If $0<x \leqslant 1, y>1$, $F(x,y) = \int_0^x\int_0^1 4uv\,\mathrm{d}u\mathrm{d}v = x^2$.
- If $x>1, 0<y \leqslant 1$, $F(x,y) = \int_0^1\int_0^y 4uv\,\mathrm{d}u\mathrm{d}v = y^2$.
- If $x>1, y>1$, $F(x,y) = \iint_S f(x,y)\,\mathrm{d}x\mathrm{d}y = 1$.

The joint CDF $F(x,y)$ is

$$F(x,y) = \begin{cases} 0, & x \leqslant 0 \text{ or } y \leqslant 0 \\ x^2y^2, & 0<x \leqslant 1, 0<y \leqslant 1 \\ x^2, & 0<x \leqslant 1, y>1 \\ y^2, & x>1, 0<y \leqslant 1 \\ 1, & x>1, y>1 \end{cases}.$$

Problem 3.12 Let the joint PDF of two random variable X and Y be given by

$$f(x,y) = \begin{cases} Cx\,\mathrm{e}^{-y}, & 0<x<y<+\infty \\ 0, & \text{otherwise} \end{cases}.$$

(1) Find the constant C.

(2) Are X and Y independent? Why?

(3) Find $f_{X|Y}(x|y)$ and $f_{Y|X}(y|x)$.

(4) Find $P(X<1|Y<2)$ and $P(X<1|Y=2)$.

Solution:

(1) See Figure 3.17.

$$\begin{aligned}1 &= \int_{-\infty}^{+\infty}\int_{-\infty}^{+\infty} f(x,y)\,dx\,dy \\ &= \int_0^{+\infty}\left(\int_0^y Cx\,e^{-y}dx\right)dy \\ &= \int_0^{+\infty}\frac{Cy^2}{2}e^{-y}dy \\ &= C.\end{aligned}$$

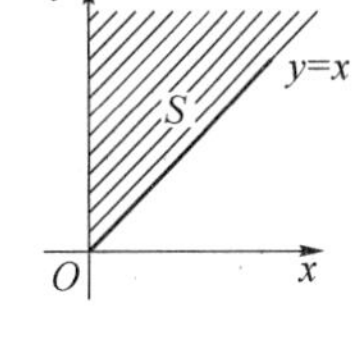

Figure 3.17

Therefore, $C=1$. The joint PDF is

$$f(x,y)=\begin{cases}x\,e^{-y}, & 0<x<y<+\infty\\ 0, & \text{otherwise}\end{cases}.$$

(2) Find the marginal PDF of X, as shown in Figure 3.18.

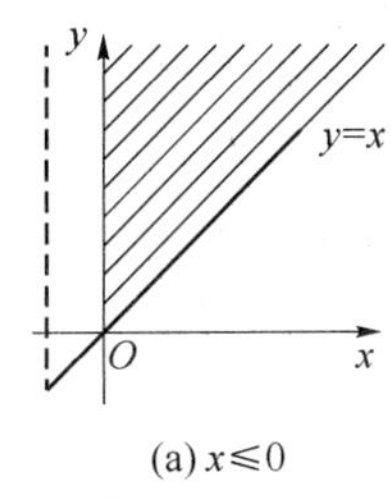

(a) $x\leq0$

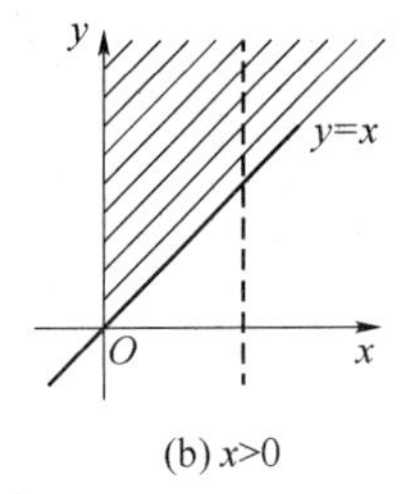

(b) $x>0$

Figure 3.18

- If $x\leq0$, $f_X(x)=0$.
- If $x>0$,

$$\begin{aligned}f_X(x) &= \int_{-\infty}^{+\infty} f(x,y)\,dy \\ &= \int_x^{+\infty} xe^{-y}dy \\ &= xe^{-x}.\end{aligned}$$

Therefore, the marginal PDF of X is

$$f_X(x)=\begin{cases}x\,e^{-x}, & x>0\\ 0, & \text{otherwise}\end{cases}.$$

Find the marginal PDF of Y, as shown in Figure 3.19.

- If $y \leqslant 0, f_Y(y)=0$.
- If $y>0$,

$$f_Y(y)=\int_{-\infty}^{+\infty} f(x,y)\,\mathrm{d}x = \int_0^y x\,\mathrm{e}^{-y}\,\mathrm{d}x = \frac{y^2}{2}\mathrm{e}^{-y}.$$

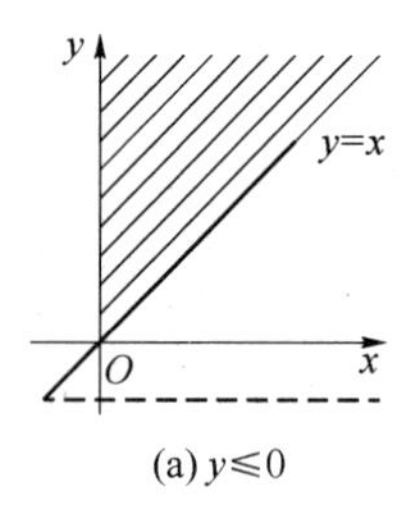

(a) $y \leqslant 0$

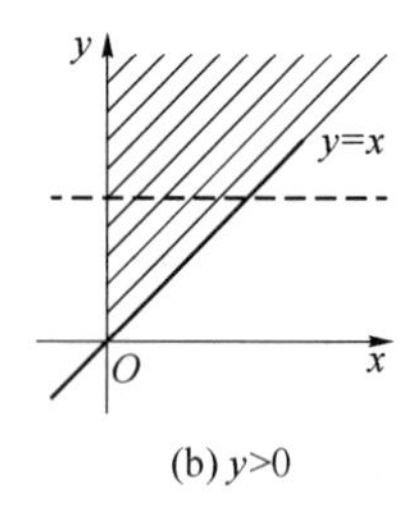

(b) $y>0$

Figure 3.19

Therefore, the marginal PDF of Y is

$$f_Y(y)=\begin{cases}\dfrac{y^2}{2}\mathrm{e}^{-y}, & y>0\\ 0, & \text{otherwise}\end{cases}.$$

In the region $\{(x,y)\mid 0<x<y\}$, $f(x,y)\neq f_X(x)f_Y(y)$, thus X and Y are not independent.

(3)

$$f_{X\mid Y}(x\mid y)=\frac{f(x,y)}{f_Y(y)}=\begin{cases}\dfrac{2x}{y^2}, & 0<x<y\\ 0, & \text{otherwise}\end{cases}$$

$$f_{Y\mid X}(y\mid x)=\frac{f(x,y)}{f_X(x)}=\begin{cases}\mathrm{e}^{x-y}, & 0<x<y\\ 0, & \text{otherwise}\end{cases}.$$

(4) Calculate $P(X<1\mid Y<2)$, as shown in Figure 3.20.

$$P(X<1\mid Y<2)=\frac{P(X<1,Y<2)}{P(Y<2)} = \frac{\int_0^1\left(\int_x^2 x\,\mathrm{e}^{-y}\mathrm{d}y\right)\mathrm{d}x}{\int_0^2 \frac{y^2}{2}\mathrm{e}^{-y}\mathrm{d}y} = \frac{\int_0^1 x(\mathrm{e}^{-x}-\mathrm{e}^{-2})\,\mathrm{d}x}{\left[\frac{1}{2}\mathrm{e}^{-y}(-y^2-2y-2)\right]_{y=0}^{y=2}} = \frac{1-\frac{1}{2\,\mathrm{e}^2}-\frac{2}{\mathrm{e}}}{1-\frac{5}{\mathrm{e}^2}} = \frac{1+4\mathrm{e}-2\,\mathrm{e}^2}{10-2\,\mathrm{e}^2}.$$

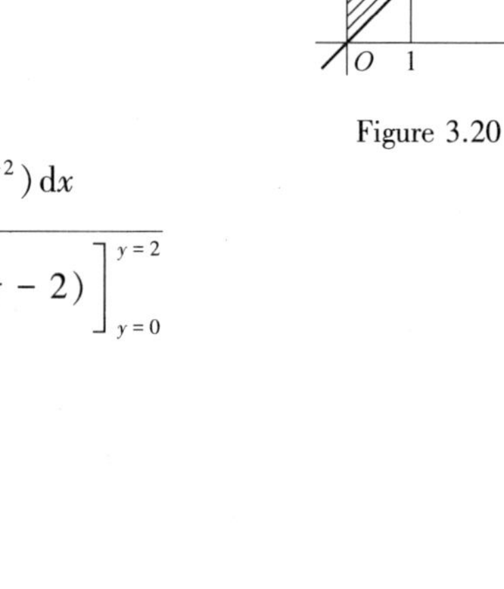

Figure 3.20

Calculate $P(X<1\mid Y=2)$. The conditional PDF of X given that $Y=2$ is

$$f_{X|Y=2}(x|y=2)=\begin{cases}\dfrac{x}{2}, & 0<x<2\\ 0, & \text{otherwise}\end{cases}$$

$$P(X<1|Y=2)=\int_0^1 \frac{x}{2}\mathrm{d}x=\frac{1}{4}.$$

Problem 3.13 The PDF of X is

$$f(x)=\begin{cases}\lambda^2 x\,\mathrm{e}^{-\lambda x}, & x>0\\ 0, & x\leqslant 0\end{cases}.$$

Y is uniformly distributed on $(0,X)$.

Find:

(1) $f_{Y|X}(y|x)$;

(2) the joint PDF of (X,Y);

(3) the PDF of Y.

 **Solution**:

(1) $f_{Y|X}(y|x)=\begin{cases}\dfrac{1}{x}, & 0<y<x\\ 0, & \text{otherwise}\end{cases}$.

(2) $f(x,y)=f_X(x)f_{Y|X}(y|x)=\begin{cases}\lambda^2\mathrm{e}^{-\lambda x}, & 0<y<x\\ 0, & \text{otherwise}\end{cases}$.

(3) See Figure 3.21.

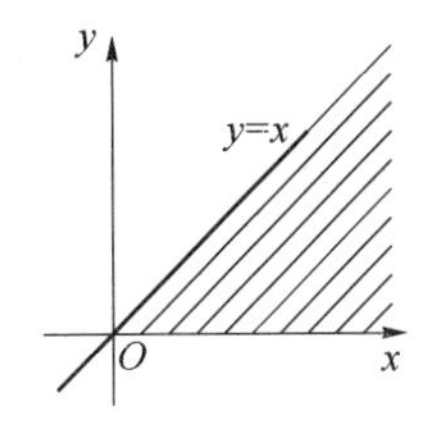

Figure 3.21

- If $y\leqslant 0$, $f_Y(y)=0$.
- If $y>0$,

$$\begin{aligned}f_Y(y)&=\int_{-\infty}^{+\infty}f(x,y)\,\mathrm{d}x\\&=\int_y^{+\infty}\lambda^2\mathrm{e}^{-\lambda x}\mathrm{d}x\\&=\lambda\,\mathrm{e}^{-\lambda y}.\end{aligned}$$

Therefore, the PDF of Y is

$$f_Y(y)=\begin{cases}\lambda\,e^{-\lambda y}, & y>0\\ 0, & \text{otherwise}\end{cases}.$$

Problem 3.14 The joint PDF of X and Y is given by

$$f(x,y)=\begin{cases}\dfrac{1}{2x^2y}, & 1\leqslant x,\dfrac{1}{x}\leqslant y\leqslant x\\ 0, & \text{otherwise}\end{cases}.$$

Are X and Y independent?

Solution:

Find the marginal PDF of X, as shown in Figure 3.22.

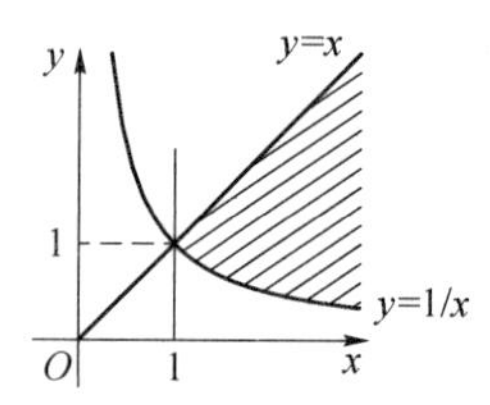

Figure 3.22

- If $x<1, f_X(x)=0$.
- If $x\geqslant 1, f_X(x)=\int_{-\infty}^{+\infty}f(x,y)\,\mathrm{d}y=\int_{1/x}^{x}\dfrac{1}{2x^2y}\mathrm{d}y=\dfrac{\ln x}{x^2}$.

Therefore, the marginal PDF of X is

$$f_X(x)=\begin{cases}\dfrac{\ln x}{x^2}, & x\geqslant 1\\ 0, & \text{otherwise}\end{cases}.$$

Find the marginal PDF of Y.

- If $y\leqslant 0, f_Y(y)=0$.
- If $0<y\leqslant 1, f_Y(y)=\int_{-\infty}^{+\infty}f(x,y)\,\mathrm{d}x=\int_{1/y}^{+\infty}\dfrac{1}{2x^2y}\mathrm{d}x=\dfrac{1}{2}$.
- If $y>1, f_Y(y)=\int_{-\infty}^{+\infty}f(x,y)\,\mathrm{d}x=\int_{y}^{+\infty}\dfrac{1}{2x^2y}\mathrm{d}x=\dfrac{1}{2y^2}$.

Therefore, the marginal PDF of Y is

$$f_Y(y)=\begin{cases}0, & y\leqslant 0\\ \dfrac{1}{2}, & 0<y\leqslant 1\\ \dfrac{1}{2y^2}, & y>1\end{cases}.$$

In the support of (X,Y), $f(x,y)\neq f_X(x)f_Y(y)$, so X and Y are not independent.

Problem 3.15 X and Y are two independent random variables. $X \sim U(0,1)$. The PDF of Y is given by

$$f(y)=\begin{cases}\frac{1}{2}e^{-\frac{y}{2}}, & y>0\\ 0, & y\leqslant 0\end{cases}.$$

(1) Determine the joint PDF of (X,Y).

(2) Find the probability that the following equation with respect to a has real roots.

$$aX^2+2aX+Y^2=0.$$

Solution:

(1) The PDF of X is $f_X(x)=\begin{cases}1, & 0<x<1\\ 0, & \text{otherwise}\end{cases}$, since X and Y are independent,

$$f(x,y)=f_X(x)f_Y(y)=\begin{cases}\frac{1}{2}e^{-\frac{y}{2}}, & 0<x<1, y>0\\ 0, & \text{otherwise}\end{cases}.$$

(2) See Figure 3.23.

$$\begin{aligned}P(\text{the equation has real roots}) &= P[(2X)^2-4Y^2\geqslant 0]\\ &=P(X\geqslant Y)\\ &=\iint_{x\geqslant y} f(x,y)\,dxdy\\ &=\int_0^1\left(\int_0^x \frac{1}{2}e^{-\frac{y}{2}}dy\right)dx\\ &=\int_0^1(1-e^{-\frac{x}{2}})dx\\ &=\frac{2}{\sqrt{e}}-1.\end{aligned}$$

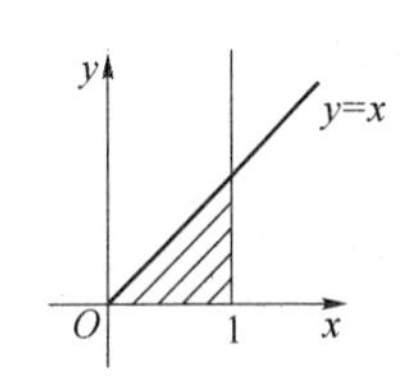

Figure 3.23

Problem 3.16 Let a needle of length L be thrown at random onto a horizontal plane ruled with parallel straight lines spaced by a distance d from each other, with $d>L$. What is the probability that the needle will intersect one of these lines?

Solution:

Let X denote the distance from the center of the needle to the nearest parallel line, and let θ denote the angle between the needle and the projection line of X, as shown in Figure 3.24.

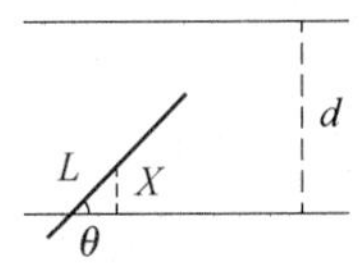

Figure 3.24

$$X\sim \mathrm{U}\left[0,\frac{d}{2}\right]\sim f_X(x)=\begin{cases}\frac{2}{d}, & 0\leqslant x\leqslant\frac{d}{2}\\ 0, & \text{otherwise}\end{cases}.$$

$$\theta\sim \mathrm{U}[0,\pi]\sim f_\theta(\theta)=\begin{cases}\frac{1}{\pi}, & 0\leqslant\theta\leqslant\pi\\ 0, & \text{otherwise}\end{cases}.$$

The joint PDF of (X,θ) is

$$f(x,\theta)=\begin{cases}\frac{2}{d\pi}, & 0\leqslant x\leqslant\frac{d}{2},0\leqslant\theta\leqslant\pi\\ 0, & \text{otherwise}\end{cases}.$$

As shown in Figure 3.25, P(the needle intersects the parallel line)

$$\begin{aligned}&=P\left(X\leqslant\frac{L}{2}\sin\theta\right)\\&=\iint_{x\leqslant\frac{L}{2}\sin\theta}f(x,\theta)\,\mathrm{d}x\mathrm{d}\theta\\&=\int_0^\pi\left(\int_0^{\frac{L}{2}\sin\theta}\frac{2}{d\pi}\mathrm{d}x\right)\mathrm{d}\theta\\&=\frac{2L}{d\pi}.\end{aligned}$$

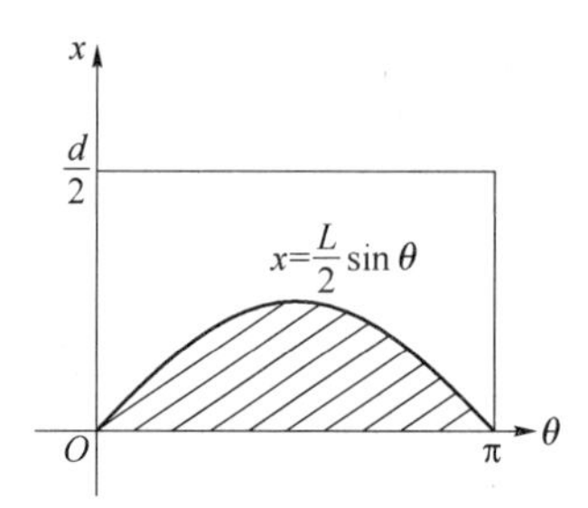

Figure 3.25

Problem 3.17 Assume that U is uniform on $[0,1]$ and that $F_X(x)$ is the CDF of a continuous random variable X. Show that $F_X^{-1}(U)$ has the same distribution as X.

Solution:

Consider the distribution of $Y=F_X^{-1}(U)$:

$$\begin{aligned}F_Y(x)&=P(Y\leqslant x)\\&=P(F_X^{-1}(U)\leqslant x)\\&=P(U\leqslant F_X(x))\\&=F_X(x).\end{aligned}$$

Since $F_X(x) \in [0,1]$ for all x, and $P(U \leqslant y) = y$ whenever $y \in [0,1]$. This shows that $F_X^{-1}(U)$ *and* X have the same distribution.

Problem 3.18 Suppose $X \sim \mathrm{P}(\lambda_1)$ and $Y \sim \mathrm{P}(\lambda_2)$ are independent. Find the probability function of $Z = X+Y$.

Solution:

The possible values of Z are $0,1,2,\cdots$.

$$\begin{aligned} P(Z=i) &= P(X+Y=i) \\ &= \sum_{k=0}^{i} P(X=k, Y=i-k) \\ &= \sum_{k=0}^{i} P(X=k) P(Y=i-k) \\ &= \sum_{k=0}^{i} \left[\frac{\lambda_1^k}{k!} \mathrm{e}^{-\lambda_1} \times \frac{\lambda_2^{i-k}}{(i-k)!} \mathrm{e}^{-\lambda_2} \right] \\ &= \frac{\mathrm{e}^{-(\lambda_1+\lambda_2)}}{i!} \sum_{k=0}^{i} \left[\frac{i! \lambda_1^k}{k!} \times \frac{\lambda_2^{i-k}}{(i-k)!} \right] \\ &= \frac{\mathrm{e}^{-(\lambda_1+\lambda_2)}}{i!} (\lambda_1 + \lambda_2)^i, i = 0,1,2,\cdots. \end{aligned}$$

Therefore, $Z \sim \mathrm{P}(\lambda_1 + \lambda_2)$.

Problem 3.19 Suppose the joint PDF of two random variables X and Y is given by

$$f(x,y) = \begin{cases} 1, & 0<x<1, 0<y<1 \\ 0, & \text{otherwise} \end{cases}.$$

Find the PDF of $Z = X+Y$.

Solution:

By the convolution formula $f_Z(z) = \int_{-\infty}^{+\infty} f(x, z-x)\,\mathrm{d}x$,

$$f(x, z-x) = \begin{cases} 1, & 0<x<1, 0<z-x<1 \\ 0, & \text{otherwise} \end{cases}.$$

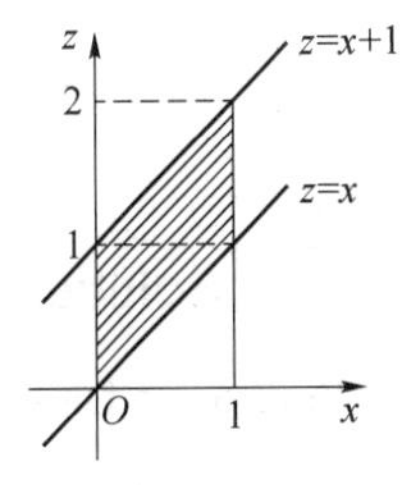

Figure 3.26

The region $\{(x,z) \mid 0<x<1, 0<z-x<1\}$ is plotted in Figure 3.26.

- If $z \leqslant 0$ or $z > 2$, $f_Z(z) = 0$.
- If $0 < z \leqslant 1$, $f_Z(z) = \int_0^z \mathrm{d}x = z$.
- If $1 < z \leqslant 2$, $f_Z(z) = \int_{z-1}^1 \mathrm{d}x = 2 - z$.

Therefore, the PDF of $Z = X+Y$ is

$$f_Z(z)=\begin{cases} z, & 0<z\leqslant 1 \\ 2-z, & 1<z\leqslant 2 \\ 0, & \text{otherwise} \end{cases}.$$

Problem 3.20 Suppose the joint PDF of two random variables X and Y is given by

$$f(x,y)=\begin{cases} 3x, & 0<x<1, 0<y<x \\ 0, & \text{otherwise} \end{cases}.$$

Find the PDF of $Z=X-Y$.

Solution 1:

By the PDF of the difference of two random variables, $f_Z(z)=\int_{-\infty}^{+\infty} f(x,x-z)\,\mathrm{d}x$,

$$f(x,x-z)=\begin{cases} 3x, & 0<x<1, 0<x-z<x \\ 0, & \text{otherwise} \end{cases}.$$

The region $\{(x,z)\mid 0<x<1, 0<x-z<x\}$ is shown in Figure 3.27.

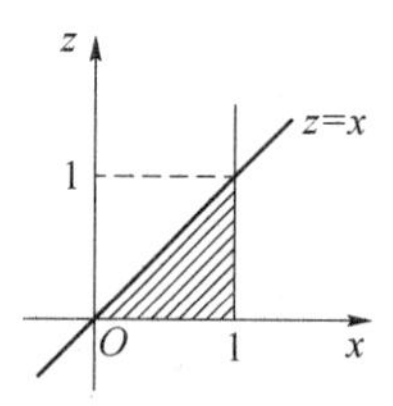

Figure 3.27

- If $z\leqslant 0$ or $z\geqslant 1$, $f_Z(z)=0$.
- If $0<z<1$, $f_Z(z)=\int_z^1 3x\mathrm{d}x=\dfrac{3(1-z^2)}{2}$.

Therefore, the PDF of $Z=X-Y$ is

$$f_Z(z)=\begin{cases} \dfrac{3(1-z^2)}{2}, & 0<z<1 \\ 0, & \text{otherwise} \end{cases}.$$

Solution 2:

Find the CDF $F_Z(z)$ of $Z=X-Y$.

$$\begin{aligned} F_Z(z) &= P(Z\leqslant z) \\ &= P(X-Y\leqslant z) \\ &= \iint_{x-y\leqslant z} f(x,y)\,\mathrm{d}x\mathrm{d}y. \end{aligned}$$

The support of (X,Y) is shown in Figure 3.28.

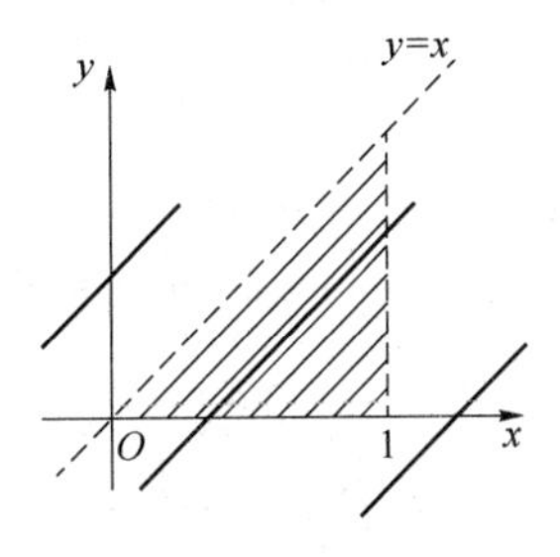

Figure 3.28

- If $z \leqslant 0, F_Z(z) = P(\varnothing) = 0$.
- If $0<z<1$,

$$
\begin{aligned}
F_Z(z) &= \iint_{x-y \leqslant z} f(x,y)\,dxdy \\
&= \int_0^z \left(\int_0^x 3x dy \right) dx + \int_z^1 \left(\int_{x-z}^x 3x dy \right) dx \\
&= \frac{1}{2}(3z - z^3).
\end{aligned}
$$

- If $z \geqslant 1$

$$
\begin{aligned}
F_Z(z) &= \iint_{x-y \leqslant z} f(x,y)\,dxdy \\
&= \int_0^1 \left(\int_0^x 3x dy \right) dx \\
&= \int_0^1 3x^2 dx \\
&= 1.
\end{aligned}
$$

Therefore, the CDF of $Z=X-Y$ is

$$
F_Z(z) = \begin{cases} 0, & z \leqslant 0 \\ \dfrac{1}{2}(3z-z^3), & 0<z<1. \\ 1, & z \geqslant 1 \end{cases}
$$

The PDF of $Z=X-Y$ is

$$
f_Z(z) = F'_Z(z) = \begin{cases} \dfrac{3}{2}(1-z^2), & 0<z<1 \\ 0, & \text{otherwise} \end{cases}.
$$

Solution 3:

Firstly, we determine the joint PDF of $Z=X-Y$ and $W=X$, then find the marginal density of Z. The joint PDF of X and Y is

$$
f(x,y) = \begin{cases} 3x, & 0<x<1, 0<y<x \\ 0, & \text{otherwise} \end{cases}.
$$

Let

$$S=\{(x,y)\mid 0<x<1,0<y<x\}$$
$$T=\{(z,w)\mid 0<w<1,0<w-z<w\}$$

then $Z=X-Y$ and $W=X$ define a one-to-one transformation from S onto T, its inverse transformation is $X=W, Y=W-Z$.

The Jacobian determinant is

$$J=\begin{vmatrix} \frac{\partial x}{\partial z} & \frac{\partial x}{\partial w} \\ \frac{\partial y}{\partial z} & \frac{\partial y}{\partial w} \end{vmatrix}=\begin{vmatrix} 0 & 1 \\ -1 & 1 \end{vmatrix}=1, |J|=1.$$

Thus, the joint PDF of Z and W is

$$f_{Z,W}(z,w)=\begin{cases} 3w, & 0<w<1,0<w-z<w \\ 0, & \text{otherwise} \end{cases}.$$

The region $\{(z,w)\mid 0<w<1,0<w-z<w\}$ is sketched in Figure 3.29.

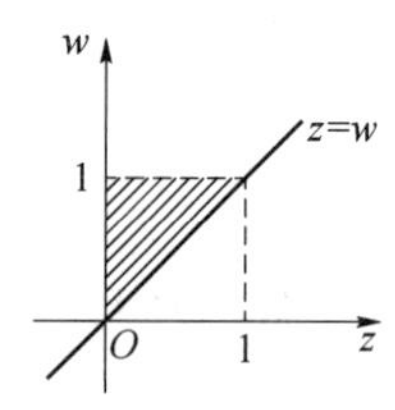

Figure 3.29

Next we determine the marginal density of Z, $f_Z(z)=\int_{-\infty}^{\infty} f_{Z,W}(z,w)\,\mathrm{d}w$.

(1) If $z\leqslant 0$ or $z\geqslant 1$, $f_Z(z)=0$.

(2) If $0<z<1$, $f_Z(z)=\int_z^1 3w\mathrm{d}w=\frac{3(1-z^2)}{2}$.

In summary, the PDF of $Z=X-Y$ is

$$f_Z(z)=\begin{cases} \frac{3(1-z^2)}{2}, & 0<z<1 \\ 0, & \text{otherwise} \end{cases}.$$

Solution 4:

Firstly, we determine the joint PDF of $Z=X-Y$ and $W=Y$, then find the marginal density of Z. The joint PDF of X and Y is

$$f(x,y)=\begin{cases} 3x, & 0<x<1,0<y<x \\ 0, & \text{otherwise} \end{cases}.$$

Let

$$S=\{(x,y)\mid 0<x<1,0<y<x\}$$

$$T=\{(z,w)\mid 0<z+w<1,0<w<z+w\}$$

then $Z=X-Y$ and $W=Y$ define a one-to-one transformation from S onto T, its inverse transformation is $X=Z+W, Y=W$.

The Jacobian determinant is

$$J=\begin{vmatrix} \frac{\partial x}{\partial z} & \frac{\partial x}{\partial w} \\ \frac{\partial y}{\partial z} & \frac{\partial y}{\partial w} \end{vmatrix}=\begin{vmatrix} 1 & 1 \\ 0 & 1 \end{vmatrix}=1, |J|=1.$$

Thus, the joint PDF of Z and W is

$$f_{Z,W}(z,w)=\begin{cases} 3(z+w), & 0<z+w<1,0<w<z+w \\ 0, & \text{otherwise} \end{cases}.$$

The region $\{(z,w)\mid 0<z+w<1,0<w<z+w\}$ is sketched in Figure 3.30.

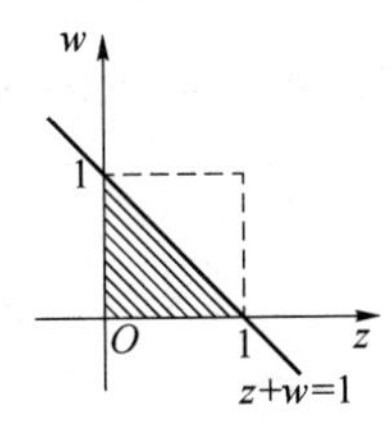

Figure 3.30

Next we determine the marginal density of $Z, f_Z(z)=\int_{-\infty}^{\infty} f_{Z,W}(z,w)\,dw$.

(1) If $z\leqslant 0$ or $z\geqslant 1, f_Z(z)=0$.

(2) If $0<z<1, f_Z(z)=\int_0^{1-z} 3(z+w)\,dw=\frac{3(1-z^2)}{2}$.

In summary, the PDF of $Z=X-Y$ is

$$f_Z(z)=\begin{cases} \frac{3(1-z^2)}{2}, & 0<z<1 \\ 0, & \text{otherwise} \end{cases}.$$

 **Solution 5:**

Firstly, we determine the joint PDF of $Z=X-Y$ and $W=X+Y$, then find the marginal density of Z.

The joint PDF of X and Y is

$$f(x,y)=\begin{cases} 3x, & 0<x<1,0<y<x \\ 0, & \text{otherwise} \end{cases}.$$

Let

$$S=\{(x,y)\mid 0<x<1,0<y<x\}$$

$$T=\left\{(z,w)\mid 0<\frac{1}{2}z+\frac{1}{2}w<1,0<-\frac{1}{2}z+\frac{1}{2}w<\frac{1}{2}z+\frac{1}{2}w\right\}$$

$$=\{(z,w)\mid 0<z+w<2, z<w, z>0\}$$

then $Z=X-Y$ and $W=X+Y$ define a one-to-one transformation from S onto T, its inverse transformation is $X=\frac{1}{2}Z+\frac{1}{2}W, Y=-\frac{1}{2}Z+\frac{1}{2}W$.

The Jacobian determinant is

$$J=\begin{vmatrix}\frac{\partial x}{\partial z} & \frac{\partial x}{\partial w}\\ \frac{\partial y}{\partial z} & \frac{\partial y}{\partial w}\end{vmatrix}=\begin{vmatrix}\frac{1}{2} & \frac{1}{2}\\ -\frac{1}{2} & \frac{1}{2}\end{vmatrix}=\frac{1}{2}, |J|=\frac{1}{2}.$$

Thus, the joint PDF of Z and W is

$$f_{Z,W}(z,w)=\begin{cases}\frac{3(z+w)}{4}, & 0<z+w<2, z<w, z>0\\ 0, & \text{otherwise}\end{cases}.$$

The region $\{(z,w)\mid 0<z+w<2, z<w, z>0\}$ is sketched in Figure 3.31.

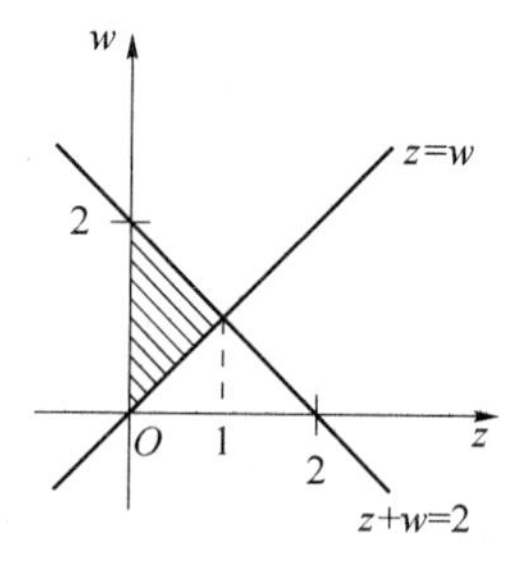

Figure 3.31

Next we determine the marginal density of Z, $f_Z(z)=\int_{-\infty}^{\infty}f_{Z,W}(z,w)\,dw$.

(1) If $z\leqslant 0$ or $z\geqslant 1$, $f_Z(z)=0$.

(2) If $0<z<1$, $f_Z(z)=\int_z^{2-z}\frac{3(z+w)}{4}dw=\frac{3(1-z^2)}{2}$.

In summary, the PDF of $Z=X-Y$ is

$$f_Z(z)=\begin{cases}\frac{3(1-z^2)}{2}, & 0<z<1\\ 0, & \text{otherwise}\end{cases}.$$

Problem 3.21 Suppose that X_1 and X_2 are IID random variables and that the PDF of each of them is as follows:

$$f(x)=\begin{cases}e^{-x}, & x>0\\ 0, & \text{otherwise}\end{cases}.$$

Find the PDF of $Y=X_1-X_2$.

Solution 1:

The joint PDF of (X_1, X_2) is as follows:

$$f(x_1, x_2) = f_{X_1}(x_1) \times f_{X_2}(x_2) = \begin{cases} e^{-(x_1+x_2)}, & x_1>0, x_2>0 \\ 0, & \text{otherwise} \end{cases}.$$

The PDF of $Y=X_1-X_2$ is

$$f_Y(y) = \int_{-\infty}^{+\infty} f(y + x_2, x_2)\,\mathrm{d}x_2$$

$$f(y+x_2, x_2) = \begin{cases} e^{-(y+2x_2)}, & y+x_2>0, x_2>0 \\ 0, & \text{otherwise} \end{cases}$$

where the region $\{(x_2, y) \mid y+x_2>0, x_2>0\}$ is sketched in Figure 3.32.

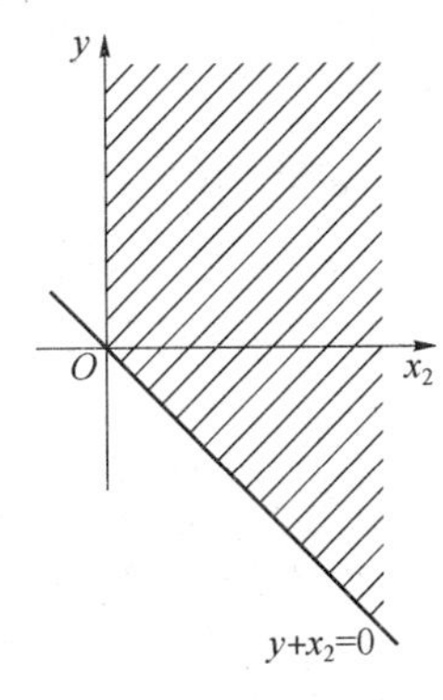

Figure 3.32

- If $y \leqslant 0$,

$$f_Y(y) = \int_{-y}^{+\infty} e^{-(y+2x_2)}\,\mathrm{d}x_2 = \frac{1}{2}e^{y}$$

- If $y>0$,

$$f_Y(y) = \int_{0}^{+\infty} e^{-(y+2x_2)}\,\mathrm{d}x_2 = \frac{1}{2}e^{-y}$$

In summary, the PDF of $Y=X_1-X_2$ is

$$f_Y(y) = \begin{cases} \frac{1}{2}e^{-y}, & y>0 \\ \frac{1}{2}e^{y}, & y \leqslant 0 \end{cases}.$$

Solution 2:

Firstly, we determine the joint PDF of $W=X_2$ and $Y=X_1-X_2$, then find the marginal density of Y.

The joint PDF of X_1 and X_2 is

$$f(x_1,x_2)=f_{X_1}(x_1)\times f_{X_2}(x_2)=\begin{cases}e^{-(x_1+x_2)}, & x_1>0,x_2>0\\0, & \text{otherwise}\end{cases}.$$

Let

$$S=\{(x_1,x_2)\mid x_1>0,x_2>0\}$$
$$T=\{(w,y)\mid w>0,w+y>0\}$$

then $W=X_2$ and $Y=X_1-X_2$ define a one-to-one transformation from S onto T, its inverse transformation is $X_1=W+Y, X_2=W$.

The Jacobian determinant is

$$J=\begin{vmatrix}\frac{\partial x_1}{\partial w} & \frac{\partial x_1}{\partial y}\\ \frac{\partial x_2}{\partial w} & \frac{\partial x_2}{\partial y}\end{vmatrix}=\begin{vmatrix}1 & 1\\1 & 0\end{vmatrix}=-1,\ |J|=1.$$

Thus, the joint PDF of W and Y is

$$f_{W,Y}(w,y)=\begin{cases}e^{-(2w+y)}, & w>0,w+y>0\\0, & \text{otherwise}\end{cases}.$$

The region $\{(w,y)\mid w>0,w+y>0\}$ is sketched in Figure 3.33.

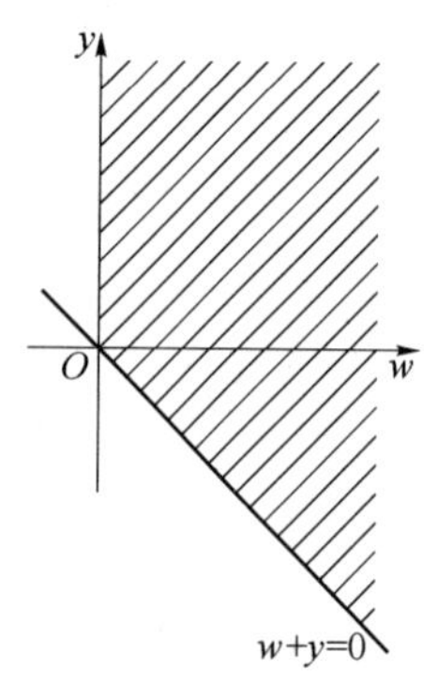

Figure 3.33

Next we determine the marginal density of Y, $f_Y(y)=\int_{-\infty}^{+\infty}f_{W,Y}(w,y)\,dw$.

- If $y\leqslant 0$,

$$f_Y(y)=\int_{-y}^{+\infty}e^{-(2w+y)}\,dw=\frac{1}{2}e^{y}.$$

- If $y>0$,

$$f_Y(y)=\int_{0}^{+\infty}e^{-(2w+y)}\,dw=\frac{1}{2}e^{-y}.$$

In summary, the PDF of $Y=X_1-X_2$ is

$$f_Y(y)=\begin{cases}\frac{1}{2}e^{-y}, & y>0\\ \frac{1}{2}e^{y}, & y\leqslant 0\end{cases}.$$

Solution 3:

Firstly, we determine the joint PDF of $W=X_1+X_2$ and $Y=X_1-X_2$, then find the marginal density of Y.

The joint PDF of X_1 and X_2 is

$$f(x_1,x_2)=f_{X_1}(x_1)\times f_{X_2}(x_2)=\begin{cases}e^{-(x_1+x_2)}, & x_1>0,x_2>0\\ 0, & \text{otherwise}\end{cases}.$$

Let

$$S=\{(x_1,x_2)\mid x_1>0,x_2>0\}$$
$$T=\{(w,y)\mid w+y>0,w-y>0\}$$

then $W=X_1+X_2$ and $Y=X_1-X_2$ define a one-to-one transformation from S onto T, its inverse transformation is $X_1=\frac{1}{2}(W+Y),X_2=\frac{1}{2}(W-Y)$.

The Jacobian determinant is

$$J=\begin{vmatrix}\frac{\partial x_1}{\partial w} & \frac{\partial x_1}{\partial y}\\ \frac{\partial x_2}{\partial w} & \frac{\partial x_2}{\partial y}\end{vmatrix}=\begin{vmatrix}\frac{1}{2} & \frac{1}{2}\\ \frac{1}{2} & -\frac{1}{2}\end{vmatrix}=-\frac{1}{2},\ |J|=\frac{1}{2}.$$

Thus, the joint PDF of W and Y is

$$f_{W,Y}(w,y)=\begin{cases}\frac{1}{2}e^{-w}, & w+y>0,w-y>0\\ 0, & \text{otherwise}\end{cases}.$$

The region $\{(w,y)\mid w+y>0,w-y>0\}$ is sketched in Figure 3.34.

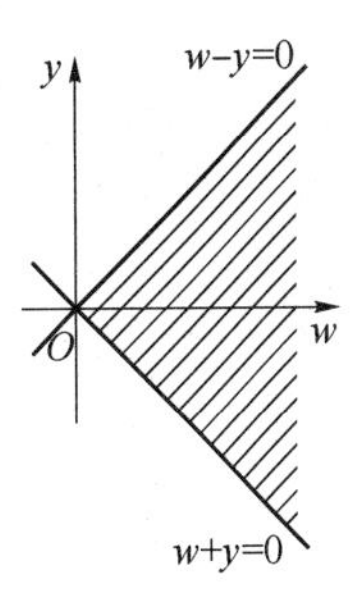

Figure 3.34

Next we determine the marginal density of Y, $f_Y(y)=\int_{-\infty}^{+\infty}f_{W,Y}(w,y)\,dw$.

- If $y\leqslant 0$,

$$f_Y(y)=\int_{-y}^{+\infty}\frac{1}{2}e^{-w}dw=\frac{1}{2}e^{y}.$$

- If $y>0$,

$$f_Y(y)=\int_{y}^{+\infty}\frac{1}{2}e^{-w}dw=\frac{1}{2}e^{-y}.$$

In summary, the PDF of $Y=X_1-X_2$ is

$$f_Y(y)=\begin{cases}\frac{1}{2}e^{-y}, & y>0\\ \frac{1}{2}e^{y}, & y\leqslant 0\end{cases}.$$

 **Solution 4**:

Firstly, we determine the CDF $F_Y(y)$ of $Y=X_1-X_2$.

$$
\begin{aligned}
F_Y(y) &= P(Y \leqslant y) \\
&= P(X_1 - X_2 \leqslant y) \\
&= \iint_{x_1 - x_2 \leqslant y} f_{X_1,X_2}(x_1, x_2)\,\mathrm{d}x_1\,\mathrm{d}x_2.
\end{aligned}
$$

Notice that the joint PDF of X_1 and X_2 is

$$
f(x_1, x_2) = f_{X_1}(x_1) \times f_{X_2}(x_2) = \begin{cases} \mathrm{e}^{-(x_1+x_2)}, & x_1>0, x_2>0 \\ 0, & \text{otherwise} \end{cases}.
$$

- If $y \leqslant 0$, as shown in Figure 3.35(a).

$$
\begin{aligned}
F_Y(y) &= \iint_{x_1 - x_2 \leqslant y} f_{X_1,X_2}(x_1, x_2)\,\mathrm{d}x_1\,\mathrm{d}x_2 \\
&= \int_{-y}^{+\infty} \left[\int_0^{x_2+y} \mathrm{e}^{-(x_1+x_2)}\,\mathrm{d}x_1 \right] \mathrm{d}x_2 \\
&= \int_{-y}^{+\infty} \mathrm{e}^{-x_2}(1 - \mathrm{e}^{-x_2-y})\,\mathrm{d}x_2 \\
&= \frac{1}{2}\mathrm{e}^{y}.
\end{aligned}
$$

- If $y>0$, as shown in Figure 3.35(b).

$$
\begin{aligned}
F_Y(y) &= \iint_{x_1 - x_2 \leqslant y} f_{X_1,X_2}(x_1, x_2)\,\mathrm{d}x_1\,\mathrm{d}x_2 \\
&= \int_{0}^{+\infty} \left[\int_0^{x_2+y} \mathrm{e}^{-(x_1+x_2)}\,\mathrm{d}x_1 \right] \mathrm{d}x_2 \\
&= \int_{0}^{+\infty} \mathrm{e}^{-x_2}(1 - \mathrm{e}^{-x_2-y})\,\mathrm{d}x_2 \\
&= 1 - \frac{1}{2}\mathrm{e}^{-y}.
\end{aligned}
$$

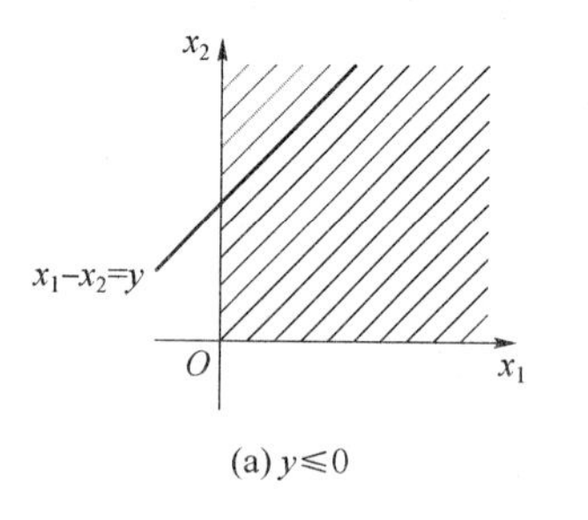

(a) $y \leqslant 0$

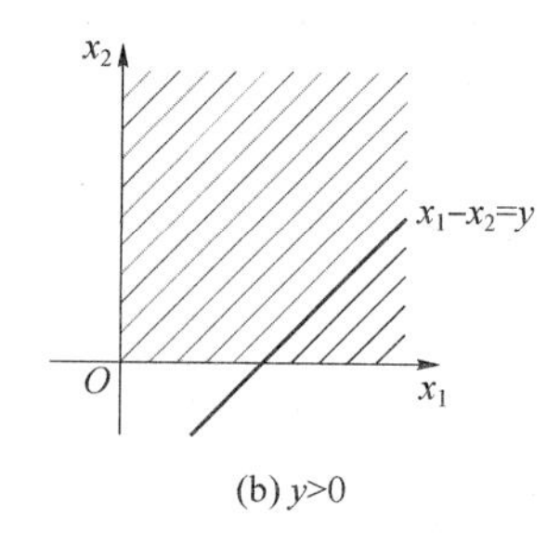

(b) $y>0$

Figure 3.35

Therefore, the CDF of $Y = X_1 - X_2$ is

$$
F_Y(y) = \begin{cases} 1 - \dfrac{1}{2}\mathrm{e}^{-y}, & y>0 \\ \dfrac{1}{2}\mathrm{e}^{y}, & y \leqslant 0 \end{cases}.
$$

The PDF of $Y = X_1 - X_2$ is

$$f_Y(y)=F'_Y(y)=\begin{cases}\frac{1}{2}e^{-y}, & y>0\\ \frac{1}{2}e^{y}, & y\leqslant 0\end{cases}.$$

Note:

- The first method is to directly use the density formula of difference.
- The second method is to introduce the auxiliary variable $W=X_2$, first find the joint PDF of (W,Y) and then find the marginal PDF of Y.
- The third method is to introduce the auxiliary variable $W=X_1+X_2$, first find the joint PDF of (W,Y) and then find the marginal PDF of Y.
- The fourth method is to find the CDF of Y first, and then find the derivative to get its PDF.

The answers obtained by the four methods are the same, and you can use them flexibly.

Problem 3.22 X and Y are two independent random variables. Their PDFs are given by

$$f_X(x)=\begin{cases}1, & 0\leqslant x\leqslant 1\\ 0, & \text{otherwise}\end{cases}, f_Y(y)=\begin{cases}e^{-y}, & y>0\\ 0, & y\leqslant 0\end{cases}.$$

Determine the PDF of $Z=X+Y$.

Solution:

By the convolution formula $f_Z(z)=\int_{-\infty}^{+\infty} f(x,z-x)\,dx$,

$$f(x,z-x)=f_X(x)f_Y(z-x)=\begin{cases}e^{x-z}, & 0\leqslant x\leqslant 1, z-x>0\\ 0, & \text{otherwise}\end{cases}.$$

The region $\{(x,z)\mid 0\leqslant x\leqslant 1, z-x>0\}$ is plotted in Figure 3.36.

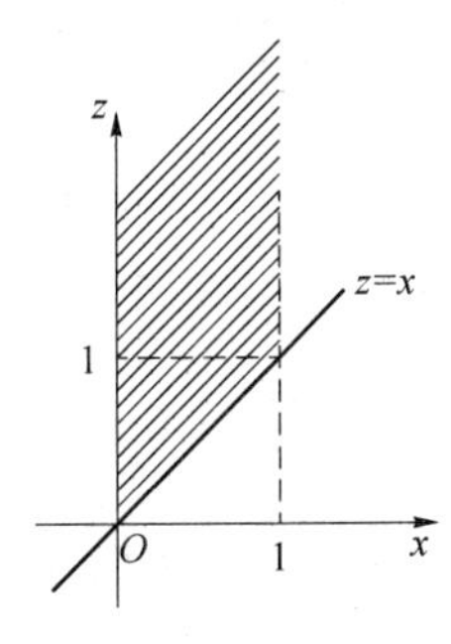

Figure 3.36

- If $z\leqslant 0$, $f_Z(z)=0$.
- If $0<z\leqslant 1$, $f_Z(z)=\int_0^z e^{x-z}dx=1-e^{-z}$.
- If $z>1$, $f_Z(z)=\int_0^1 e^{x-z}dx=e^{1-z}-e^{-z}$.

Therefore, the PDF of $Z=X+Y$ is

$$f_Z(z)=\begin{cases}0, & z\leqslant 0\\ 1-e^{-z}, & 0<z\leqslant 1\\ e^{1-z}-e^{-z}, & z>1\end{cases}.$$

Problem 3.23 Suppose the joint PDF of two random variable X and Y is given by

$$f(x,y)=\begin{cases}e^{-(x+y)}, & x>0, y>0\\ 0, & \text{otherwise}\end{cases}.$$

Determine the PDF of $Z=|X-Y|$.

 **Solution 1:**

The PDF of $W=X-Y$ is

$$f_W(w)=\int_{-\infty}^{+\infty}f(w+y,y)\,dy$$

$$f(w+y,y)=\begin{cases}e^{-(w+2y)}, & w+y>0, y>0\\ 0, & \text{otherwise}\end{cases}$$

where the region $\{(y,w)\mid w+y>0, y>0\}$ is sketched in Figure 3.37.

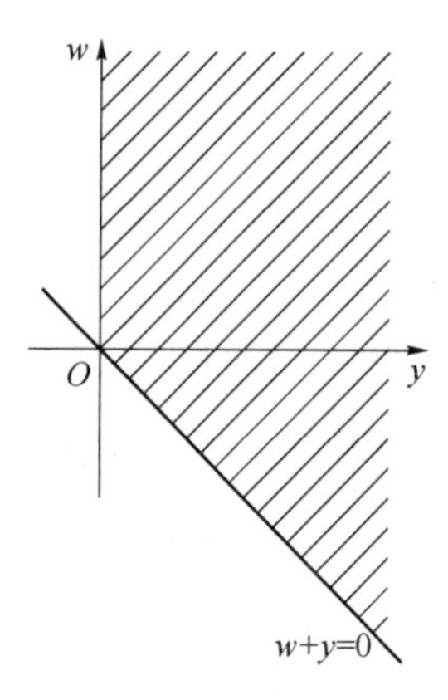

Figure 3.37

- If $w\leqslant 0$,

$$f_W(w)=\int_{-w}^{+\infty}e^{-(w+2y)}\,dy=\frac{1}{2}e^{w}.$$

- If $w>0$,

$$f_W(w)=\int_{0}^{+\infty}e^{-(w+2y)}\,dy=\frac{1}{2}e^{-w}.$$

In summary, the PDF of $W=X-Y$ is

$$f_W(w)=\begin{cases}\frac{1}{2}e^{-w}, & w>0\\ \frac{1}{2}e^{w}, & w\leqslant 0\end{cases}.$$

Find the CDF of $Z=|X-Y|=|W|$.

$$F_Z(z)=P(Z\leqslant z)=P(|W|\leqslant z)$$

(1) If $z\leqslant 0, F_Z(z)=P(\varnothing)=0$.

(2) If $z>0$,

$$\begin{aligned}F_Z(z)&=P(-z\leqslant W\leqslant z)\\&=\int_{-z}^{z} f_W(w)\,dw\\&=\int_{-z}^{0}\frac{1}{2}e^{w}dw+\int_{0}^{z}\frac{1}{2}e^{-w}dw\\&=1-e^{-z}.\end{aligned}$$

Therefore,

$$F_Z(z)=\begin{cases}1-e^{-z}, & z>0\\ 0, & z\leqslant 0\end{cases}.$$

The PDF of Z is

$$f_Z(z)=F_Z'(z)=\begin{cases}e^{-z}, & z>0\\ 0, & z\leqslant 0\end{cases}.$$

 **Solution 2:**

Firstly, we determine the CDF $F_Z(z)$ of $Z=|X-Y|$.

$$F_Z(z)=P(Z\leqslant z)=P(|X-Y|\leqslant z)$$

(1) If $z\leqslant 0, F_Z(z)=P(\varnothing)=0$.

(2) If $z>0$,

$$\begin{aligned}F_Z(z)&=P(|X-Y|\leqslant z)\\&=\iint_{|x-y|\leqslant z} f_{X,Y}(x,y)\,dxdy.\end{aligned}$$

Notice that the joint PDF of X and Y is

$$f(x,y)=\begin{cases}e^{-(x+y)}, & x>0, y>0\\ 0, & \text{otherwise}\end{cases}.$$

The support is $S=\{(x,y)\mid x>0, y>0\}$, as shown in Figure 3.38.

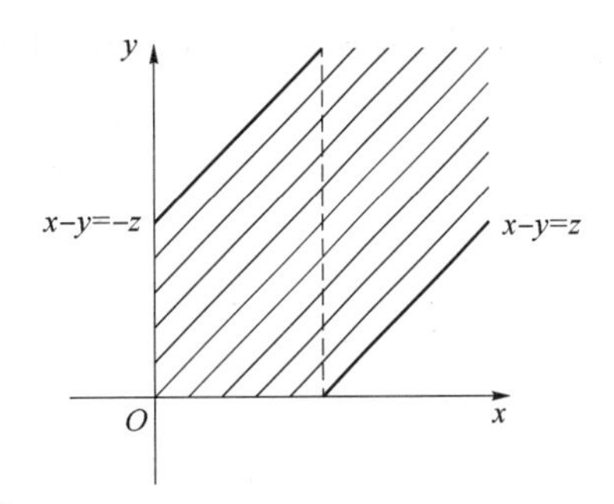

Figure 3.38

$$\begin{aligned}F_Z(z)&=\iint_{|x-y|\leqslant z\cap S} e^{-(x+y)}dxdy\\&=\int_0^z\left[\int_0^{x+z}e^{-(x+y)}dy\right]dx+\int_z^{+\infty}\left[\int_{x-z}^{x+z}e^{-(x+y)}dy\right]dx\\&=\int_0^z e^{-x}[1-e^{-(x+z)}]dx+\int_z^{+\infty}e^{-x}[e^{-(x-z)}-e^{-(x+z)}]dx\end{aligned}$$

$$= \int_0^z [e^{-x} - e^{-(2x+z)}]dx + \int_z^{+\infty} [e^{-(2x-z)} - e^{-(2x+z)}]dx$$

$$= \left[- e^{-x} + \frac{1}{2} e^{-(2x+z)} \right]_{x=0}^{x=z} + \left[- \frac{1}{2} e^{-(2x-z)} + \frac{1}{2} e^{-(2x+z)} \right]_{x=z}^{x=+\infty}$$

$$= 1 - e^{-z}.$$

Therefore, the CDF of $Z=|X-Y|$ is

$$F_Z(z)=\begin{cases}1-e^{-z}, & z>0\\0, & z\leqslant 0\end{cases}.$$

The PDF of $Z=|X-Y|$ is

$$f_Z(z)=F'_Z(z)=\begin{cases}e^{-z}, & z>0\\0, & z\leqslant 0\end{cases}.$$

 **Solution 3**:

Firstly, we determine the CDF $F_Z(z)$ of $Z=|X-Y|$.

$$F_Z(z)=P(Z\leqslant z)=P(|X-Y|\leqslant z)$$

(1) If $z\leqslant 0, F_Z(z)=P(\varnothing)=0$.

(2) If $z>0$,

$$\begin{aligned}F_Z(z)&=P(|X-Y|\leqslant z)\\&=P(-z\leqslant X-Y\leqslant z)\\&=P(X-Y\leqslant z)-P(X-Y\leqslant -z)\\&=P(X-Y\leqslant z)-P(Y-X\leqslant -z)\\&=2P(X-Y\leqslant z)-1\end{aligned}$$

It is easy to show that X and Y follow the exponential distribution E(1), and X and Y are independent. In the above, we have used the fact that $X-Y$ and $Y-X$ have the same distribution.

We'll apply the total probability to get

$$\begin{aligned}P(X - Y \leqslant z) &= P(X \leqslant Y + z)\\&= \int_0^{+\infty} P(X \leqslant y + z) f_Y(y)dy\\&= \int_0^{+\infty} F_X(y + z) f_Y(y)dy\\&= \int_0^{+\infty} [1 - e^{-(y+z)}] e^{-y}dy\\&= 1 - \frac{1}{2} e^{-z}.\end{aligned}$$

Thus, $F_Z(z)=1-e^{-z}, z>0$.

Therefore, the CDF of $Z=|X-Y|$ is

$$F_Z(z)=\begin{cases}1-e^{-z}, & z>0\\0, & z\leqslant 0\end{cases}.$$

The PDF of $Z=|X-Y|$ is

$$f_Z(z)=F'_Z(z)=\begin{cases} e^{-z}, & z>0 \\ 0, & z\leqslant 0 \end{cases}.$$

Problem 3.24 The joint PF of (X,Y) is given in the following table

	$Y=-1$	$Y=1$	$Y=2$
$X=-1$	$\frac{5}{20}$	$\frac{2}{20}$	$\frac{6}{20}$
$X=2$	$\frac{3}{20}$	$\frac{3}{20}$	$\frac{1}{20}$

Find the PF of

(1) $Z_1=X+Y$; (2) $Z_2=XY$; (3) $Z_3=\frac{X}{Y}$; (4) $Z_4=\max\{X,Y\}$.

 **Solution**:

(1) $Z_1=X+Y$ takes $-2,0,1,3$, or 4.

$$P(Z_1=-2)=P(X+Y=-2)=P(X=-1,Y=-1)=\frac{5}{20};$$

$$P(Z_1=0)=P(X+Y=0)=P(X=-1,Y=1)=\frac{2}{20};$$

$$P(Z_1=1)=P(X+Y=1)=P(X=2,Y=-1)+P(X=-1,Y=2)=\frac{9}{20};$$

$$P(Z_1=3)=P(X+Y=3)=P(X=2,Y=1)=\frac{3}{20};$$

$$P(Z_1=4)=P(X+Y=4)=P(X=2,Y=2)=\frac{1}{20}.$$

The PF of $Z_1=X+Y$ is

Z_1	-2	0	1	3	4
P	$\frac{5}{20}$	$\frac{2}{20}$	$\frac{9}{20}$	$\frac{3}{20}$	$\frac{1}{20}$

(2) $Z_2=XY$ takes $-1,-2,1,2$, or 4.

$$P(Z_2=-1)=P(XY=-1)=P(X=-1,Y=1)=\frac{2}{20};$$

$$P(Z_2=-2)=P(XY=-2)=P(X=-1,Y=2)+P(X=2,Y=-1)=\frac{6}{20}+\frac{3}{20}=\frac{9}{20};$$

$$P(Z_2=1)=P(XY=1)=P(X=-1,Y=-1)=\frac{5}{20};$$

$P(Z_2=2)=P(XY=2)=P(X=2,Y=1)=\frac{3}{20}$;

$P(Z_2=4)=P(XY=4)=P(X=2,Y=2)=\frac{1}{20}$.

The PF of $Z_2=XY$ is

Z_2	-1	-2	1	2	4
P	$\frac{2}{20}$	$\frac{9}{20}$	$\frac{5}{20}$	$\frac{3}{20}$	$\frac{1}{20}$

(3) $Z_3=\frac{X}{Y}$ takes $-2,-1,-\frac{1}{2},1$, or 2.

$P(Z_3=-2)=P\left(\frac{X}{Y}=-2\right)=P(X=2,Y=-1)=\frac{3}{20}$;

$P(Z_3=-1)=P\left(\frac{X}{Y}=-1\right)=P(X=-1,Y=1)=\frac{2}{20}$;

$P\left(Z_3=-\frac{1}{2}\right)=P\left(\frac{X}{Y}=-\frac{1}{2}\right)=P(X=-1,Y=2)=\frac{6}{20}$;

$P(Z_3=1)=P\left(\frac{X}{Y}=1\right)=P(X=-1,Y=-1)+P(X=2,Y=2)=\frac{5}{20}+\frac{1}{20}=\frac{6}{20}$;

$P(Z_3=2)=P\left(\frac{X}{Y}=2\right)=P(X=2,Y=1)=\frac{3}{20}$.

The PF of $Z_3=\frac{X}{Y}$ is

Z_3	-2	-1	$-\frac{1}{2}$	1	2
P	$\frac{3}{20}$	$\frac{2}{20}$	$\frac{6}{20}$	$\frac{6}{20}$	$\frac{3}{20}$

(4) $Z_4=\max\{X,Y\}$ takes $-1,1$, or 2.

$P(Z_4=-1)=P(\max\{X,Y\}=-1)=P(X=-1,Y=-1)=\frac{5}{20}$;

$P(Z_4=1)=P(\max\{X,Y\}=1)=P(X=-1,Y=1)=\frac{2}{20}$;

$P(Z_4=2)=P(\max\{X,Y\}=2)=P(X=-1,Y=2)+P(X=2,Y=-1)+P(X=2,Y=1)+P(X=2,Y=2)=\frac{6}{20}+\frac{3}{20}+\frac{3}{20}+\frac{1}{20}=\frac{13}{20}$.

The PF of $Z_4=\max\{X,Y\}$ is

Z_4	-1	1	2
P	$\frac{5}{20}$	$\frac{2}{20}$	$\frac{13}{20}$

Problem 3.25 Suppose that (X,Y) follows the uniform distribution on the region $D=\{(x,y)\mid 0<x<a, 0<y<a\}$.

(1) Find the probability density function of $Z=\frac{X}{Y}$.

(2) Find the probability density function of $M=\max\{X,Y\}$.

Solution:

(1) The joint PDF of (X,Y) is $f(x,y)=\begin{cases}\frac{1}{a^2}, & 0<x<a, 0<y<a\\ 0, & \text{otherwise}\end{cases}$. By the PDF for the quotient, the PDF of $Z=\frac{X}{Y}$ is

$$f_Z(z)=\int_{-\infty}^{+\infty} f(zy,y)\,|y|\,\mathrm{d}y,$$

where

$$f(zy,y)=\begin{cases}\frac{1}{a^2}, & 0<zy<a, 0<y<a\\ 0, & \text{otherwise}\end{cases}.$$

The region $\{(y,z)\mid 0<zy<a, 0<y<a\}$ is shown in Figure 3.39.

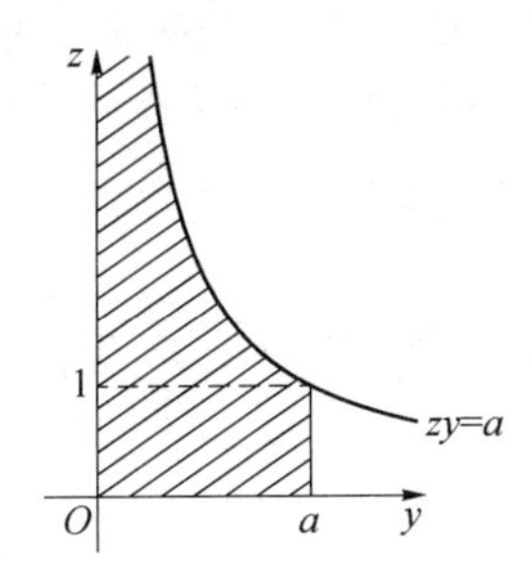

Figure 3.39

- If $z\leqslant 0$, $f_Z(z)=0$.
- If $0<z\leqslant 1$, $f_Z(z)=\int_0^a \frac{1}{a^2}y\,\mathrm{d}y=\frac{1}{2}$.
- If $z>1$, $f_Z(z)=\int_0^{a/z} \frac{1}{a^2}y\,\mathrm{d}y=\frac{1}{2z^2}$.

Therefore, the PDF of $Z=\frac{X}{Y}$ is $f_Z(z)=\begin{cases}0, & z\leqslant 0\\ \frac{1}{2}, & 0<z\leqslant 1\\ \frac{1}{2z^2}, & z>1\end{cases}$.

(2) Find the marginal PDF of X, as shown in Figure 3.40.

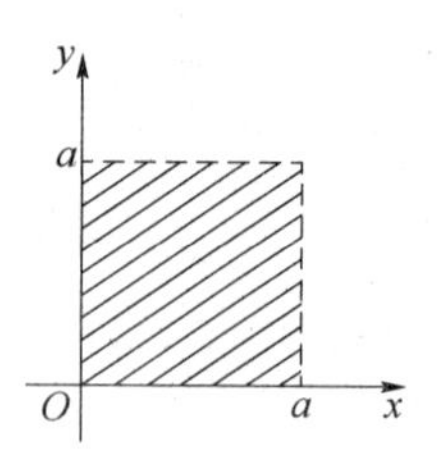

Figure 3.40

- If $x\leqslant 0$ or $x\geqslant a$, $f_X(x)=0$.
- If $0<x<a$, $f_X(x)=\int_0^a \frac{1}{a^2}\mathrm{d}y=\frac{1}{a}$.

Therefore, the PDF of X is

$$f_X(x)=\begin{cases}\frac{1}{a}, & 0<x<a \\ 0, & \text{otherwise}\end{cases}.$$

Similarly, the PDF of Y is

$$f_Y(y)=\begin{cases}\frac{1}{a}, & 0<y<a \\ 0, & \text{otherwise}\end{cases}.$$

Since $f(x,y)=f_X(x)f_Y(y)$ holds in the xOy plane, X and Y are independent.
$F_Z(z)=P(\max\{X,Y\}\leqslant z)=P(X\leqslant z, Y\leqslant z)=F_X(z)F_Y(z)$, where

$$F_X(x)=\int_{-\infty}^{x} f_X(t)\,\mathrm{d}t=\begin{cases}0, & x\leqslant 0 \\ \frac{x}{a}, & 0<x<a \\ 1, & x\geqslant a\end{cases},\quad F_Y(y)=\int_{-\infty}^{y} f_Y(t)\,\mathrm{d}t=\begin{cases}0, & y\leqslant 0 \\ \frac{y}{a}, & 0<y<a \\ 1, & y\geqslant a\end{cases}.$$

Thus, the CDF of $Z=\max\{X,Y\}$ is $F_Z(z)=F_X(z)F_Y(z)=\begin{cases}0, & z\leqslant 0 \\ \left(\frac{z}{a}\right)^2, & 0<z<a \\ 1, & z\geqslant a\end{cases}.$

The PDF of $Z=\max\{X,Y\}$ is $f_Z(z)=F_Z'(z)=\begin{cases}\frac{2z}{a^2}, & 0<z<a \\ 0, & \text{otherwise}\end{cases}.$

Note: the PDF for the quotient of two random variables can be derived as follows.

Let X and Y be two continuous random variables with joint PDF $f(x,y)$ and let $Z=\frac{X}{Y}$. The PDF of Z is

$$f_Z(z)=\int_{-\infty}^{+\infty} f(zy,y)\,|y|\,\mathrm{d}y.$$

Method 1: The joint PDF can be calculated by transformation, and then the marginal PDF can be calculated.

$$\begin{cases} Z=r_1(X,Y)=\dfrac{X}{Y} \\ W=r_2(X,Y)=Y \end{cases}$$

$$S=\{(x,y)\mid y\neq 0\}$$

$$T=\{(z,w)\mid w\neq 0\}$$

Then $Z=\dfrac{X}{Y}$ and $W=Y$ define a one-to-one transformation from S onto T, its inverse transformation is

$$\begin{cases} X=ZW \\ Y=W \end{cases}$$

The Jacobian determinant is

$$J=\begin{vmatrix} \dfrac{\partial x}{\partial z} & \dfrac{\partial x}{\partial w} \\ \dfrac{\partial y}{\partial z} & \dfrac{\partial y}{\partial w} \end{vmatrix}=\begin{vmatrix} w & z \\ 0 & 1 \end{vmatrix}=w,\ |J|=|w|.$$

Thus, the joint PDF of Z and W is

$$f_{Z,W}(z,w)=f_{X,Y}(zw,w)\,|w|.$$

The marginal PDF of Z is

$$\begin{aligned} f_Z(z) &= \int_{-\infty}^{+\infty} f_{Z,W}(z,w)\,\mathrm{d}w \\ &= \int_{-\infty}^{+\infty} f_{X,Y}(zw,w)\,|w|\,\mathrm{d}w \quad (\text{let } w=y) \\ &= \int_{-\infty}^{+\infty} f_{X,Y}(zy,y)\,|y|\,\mathrm{d}y. \end{aligned}$$

Method 2: The PDF can be obtained by first finding the CDF and then deriving it. The CDF of Z is

$$\begin{aligned} F_Z(z) &= P(Z\leqslant z) \\ &= P\left(\frac{X}{Y}\leqslant z\right) \\ &= P\left(\frac{X}{Y}\leqslant z, Y>0\right)+P\left(\frac{X}{Y}\leqslant z, Y<0\right) \\ &= P(X\leqslant zY, Y>0)+P(X\geqslant zY, Y<0) \\ &= \int_0^{+\infty}\left[\int_{-\infty}^{zy} f(x,y)\,\mathrm{d}x\right]\mathrm{d}y+\int_{-\infty}^{0}\left[\int_{zy}^{+\infty} f(x,y)\,\mathrm{d}x\right]\mathrm{d}y. \end{aligned}$$

Take derivative with respect to z to get the PDF of Z,

$$
\begin{aligned}
f_Z(z) &= F_Z'(z) \\
&= \int_0^{+\infty} [f(zy,y)y]\mathrm{d}y + \int_{-\infty}^0 [-f(zy,y)y]\mathrm{d}y \\
&= \int_0^{+\infty} f(zy,y)\,|y|\,\mathrm{d}y + \int_{-\infty}^0 f(zy,y)\,|y|\,\mathrm{d}y \\
&= \int_{-\infty}^{+\infty} f(zy,y)\,|y|\,\mathrm{d}y.
\end{aligned}
$$

Problem 3.26 Suppose that X_1, X_2, X_3, X_4 are IID and

$$P(X_i=0)=0.6, P(X_i=1)=0.4, i=1,2,3,4.$$

(1) Find the PF of $X=\begin{vmatrix} X_1 & X_2 \\ X_3 & X_4 \end{vmatrix}$.

(2) Find the probability that the following system has exactly one zero solution $x_1=x_2=0$

$$\begin{cases} X_1x_1+X_2x_2=0 \\ X_3x_1+X_4x_2=0 \end{cases}.$$

 **Solution:**

(1) The determinant $X=X_1X_4-X_2X_3$ takes $-1, 0,$ or 1.

$$
\begin{aligned}
P(X=-1) &= P(X_1X_4=0, X_2X_3=1) \\
&= P(X_1X_4=0)P(X_2X_3=1) \\
&= P(X_1=0 \cup X_4=0)P(X_2=1, X_3=1) \\
&= (0.6+0.6-0.6\times0.6)\times0.4\times0.4 \\
&= 0.134\,4.
\end{aligned}
$$

$$
\begin{aligned}
P(X=0) &= P[(X_1X_4=0, X_2X_3=0)\cup(X_1X_4=1, X_2X_3=1)] \\
&= P(X_1X_4=0, X_2X_3=0)+P(X_1X_4=1, X_2X_3=1) \\
&= P(X_1X_4=0)P(X_2X_3=0)+P(X_1X_4=1)P(X_2X_3=1) \\
&= (0.6+0.6-0.6\times0.6)^2+(0.4\times0.4)^2=0.731\,2
\end{aligned}
$$

$$
\begin{aligned}
P(X=1) &= 1-P(X=0)-P(X=-1) \\
&= 0.134\,4.
\end{aligned}
$$

Therefore, the PF of X is

X	-1	0	1
P	0.134 4	0.731 2	0.134 4

(2) This is a system of homogeneous linear equations. It has only zero solution if and only if the determinant of the coefficient matrix is not zero.

$$P(X\neq0)=1-P(X=0)=1-0.731\,2=0.268\,8.$$

Problem 3.27 Suppose that a point in the xy-plane is chosen at random from the interior of a circle for which the equation is $x^2+y^2=1$; and suppose that the probability that the point will belong to each region inside the circle is proportional to the area of that region. Let Z denote a random variable representing the distance from the center of the circle to the point. Find and sketch the CDF of Z.

Solution 1:

$P(Z \leqslant z)$ is the probability that Z lies within a circle of radius z centered at the origin. This probability is

$$\frac{\text{area of circle of radius } z}{\text{area of circle of radius } 1}=z^2, \quad 0 \leqslant z \leqslant 1.$$

The CDF is plotted in Figure 3.41.

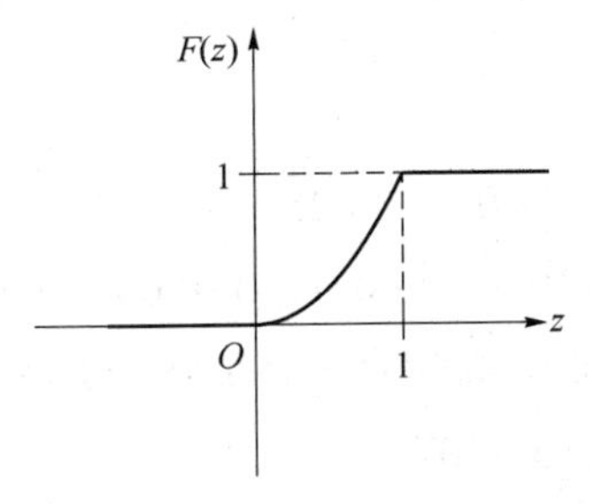

Figure 3.41

Solution 2:

(X,Y) is uniformly distributed in the interior of a circle. The joint PDF of (X,Y) is

$$f(x,y)=\begin{cases}\dfrac{1}{\pi}, & x^2+y^2 \leqslant 1 \\ 0, & \text{otherwise}\end{cases}.$$

Find the CDF of $Z=\sqrt{X^2+Y^2}$.

- If $z<0, F_Z(z)=P(\varnothing)=0$.
- If $0 \leqslant z \leqslant 1$,

$$\begin{aligned} F_Z(z) &= P(\sqrt{X^2+Y^2} \leqslant z) \\ &= \iint_{x^2+y^2 \leqslant z} f(x,y)\,\mathrm{d}x\mathrm{d}y \\ &= \frac{1}{\pi}\iint_{x^2+y^2 \leqslant z} \mathrm{d}x\mathrm{d}y \\ &= z^2. \end{aligned}$$

- If $z>1, F_Z(z)=1$.

Therefore, the CDF of Z is

$$F_Z(z)=\begin{cases}0, & z<0\\ z^2, & 0\leqslant z\leqslant 1\\ 1, & z>1\end{cases}.$$

Problem 3.28 The joint PF of (X,Y) is given in the table below.

	$Y=0$	$Y=1$	$Y=2$	$Y=3$
$X=0$	0.08	0.07	0.04	0
$X=1$	0.06	0.15	0.05	0.04
$X=2$	0.05	0.04	0.10	0.06
$X=3$	0	0.03	0.04	0.07
$X=4$	0	0.01	0.05	0.06

Determine each of the following probabilities:

(1) $P(X=2)$; (2) $P(Y\geqslant 2)$; (3) $P(X\leqslant 2, Y\leqslant 2)$; (4) $P(X=Y)$.

Solution:

(1) $P(X=2)=0.05+0.04+0.10+0.06=0.25$.

(2) $P(Y\geqslant 2)=0.04+0.05+0.10+0.04+0.05+0+0.04+0.06+0.07+0.06=0.51$.

(3) $P(X\leqslant 2, Y\leqslant 2)=0.08+0.07+0.04+0.06+0.15+0.05+0.05+0.04+0.10=0.64$.

(4) $P(X=Y)=0.08+0.15+0.10+0.07=0.40$.

Problem 3.29 Suppose that X and Y have a discrete joint distribution for which the joint PF is defined as follows:

$$f(x,y)=\begin{cases}c\,|x+y|, & x=-2,-1,0,1,2, \text{and } y=-2,-1,0,1,2\\ 0, & \text{otherwise}\end{cases}.$$

Determine:

(1) the value of the constant c;

(2) $P(X=0, Y=-2)$;

(3) $P(X=1)$;

(4) $P(|X-Y|\leqslant 1)$.

Solution:

(1) If we sum $f(x,y)$ over the 25 possible pairs of values (x,y), we obtain $40c$. Since this sum must be equal to 1, it follows that $c=\dfrac{1}{40}$. The joint PF of (X,Y) is

$$f(x,y)=\begin{cases}\dfrac{1}{40}|x+y|, & x=-2,-1,0,1,2, \text{and } y=-2,-1,0,1,2\\ 0, & \text{otherwise}\end{cases}.$$

(2) $P(X=0,Y=-2)=\frac{1}{40}\times 2=\frac{1}{20}$.

(3) $P(X=1)=\sum_{y=-2}^{2} f(1,y)=\frac{7}{40}$.

(4) The answer is found by summing $f(x,y)$ over the following pairs: $(-2,-2)$, $(-2,-1)$, $(-1,-2)$, $(-1,-1)$, $(-1,0)$, $(0,-1)$, $(0,0)$, $(0,1)$, $(1,0)$, $(1,1)$, $(1,2)$, $(2,1)$, and $(2,2)$. The sum is 0.7.

Problem 3.30 Suppose that X and Y have a continuous joint distribution for which the joint PDF is defined as follows:

$$f(x,y)=\begin{cases} cy^2, & 0\leqslant x\leqslant 2 \text{ and } 0\leqslant y\leqslant 1 \\ 0, & \text{otherwise} \end{cases}.$$

Determine:

(1) the value of the constant c;

(2) $P(X+Y>2)$;

(3) $P\left(Y<\frac{1}{2}\right)$;

(4) $P(X\leqslant 1)$;

(5) $P(X=3Y)$.

 **Solution:**

(1) $\int_{-\infty}^{+\infty}\int_{-\infty}^{+\infty} f(x,y)\,dxdy=\int_0^1\left(\int_0^2 c\,y^2 dx\right)dy=\frac{2c}{3}$. Since the value of this integral must be 1, it follows that $c=\frac{3}{2}$. The joint PDF is

$$f(x,y)=\begin{cases} \frac{3}{2}y^2, & 0\leqslant x\leqslant 2 \text{ and } 0\leqslant y\leqslant 1, \\ 0, & \text{otherwise} \end{cases}.$$

(2) The region D over which to integrate is shaded in Figure 3.42.

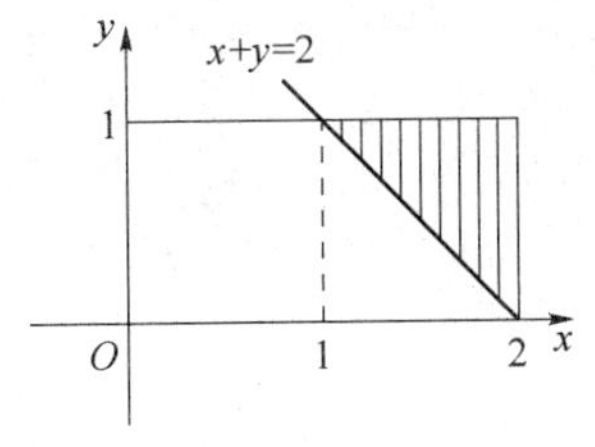

Figure 3.42

$$P(X+Y>2)=\iint_D f(x,y)\,dxdy=\int_1^2\left(\int_{2-x}^{1}\frac{3}{2}y^2 dy\right)dx=\frac{3}{8}.$$

(3) The region D over which to integrate is shaded in Figure 3.43.

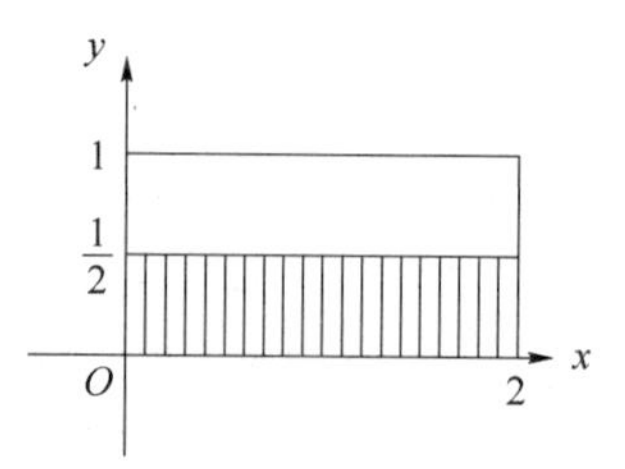

Figure 3.43

$$P\left(Y<\frac{1}{2}\right)=\iint_{D} f(x,y)\,\mathrm{d}x\mathrm{d}y=\int_{0}^{2}\left(\int_{0}^{\frac{1}{2}} \frac{3}{2} y^{2}\mathrm{d}y\right)\mathrm{d}x=\frac{1}{8}.$$

(4) The region D over which to integrate is shaded in Figure 3.44.

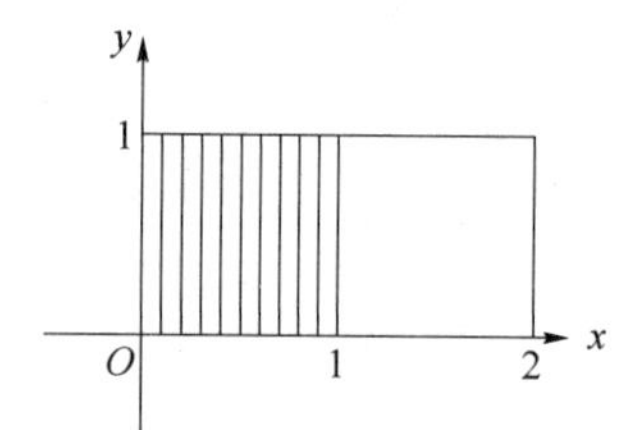

Figure 3.44

$$P(X\leqslant 1)=\iint_{D} f(x,y)\,\mathrm{d}x\mathrm{d}y=\int_{0}^{1}\left(\int_{0}^{1} \frac{3}{2} y^{2}\mathrm{d}y\right)\mathrm{d}x=\frac{1}{2}.$$

(5) The probability that (X,Y) will lie on the line $x=3y$ is 0 for every continuous joint distribution. So $P(X=3Y)=0$.

Problem 3.31 Suppose that the joint PDF of two random variables X and Y is as follows:

$$f(x,y)=\begin{cases} c(x^2+y), & 0\leqslant y\leqslant 1-x^2 \\ 0, & \text{otherwise} \end{cases}.$$

Determine:

(1) the value of the constant c;

(2) $P\left(0\leqslant X\leqslant \frac{1}{2}\right)$;

(3) $P(Y\leqslant X+1)$;

(4) $P(Y=X^2)$.

Solution:

(1) $\int_{-\infty}^{+\infty}\int_{-\infty}^{+\infty} f(x,y)\,\mathrm{d}x\mathrm{d}y=\int_{-1}^{1}\left(\int_{0}^{1-x^2} c(x^2+y)\,\mathrm{d}y\right)\mathrm{d}x=\frac{4c}{5}$. Since the value of this integral must be 1, it follows that $c=\frac{5}{4}$. The joint PDF is

$$f(x,y)=\begin{cases}\dfrac{5}{4}(x^2+y), & 0\leqslant y\leqslant 1-x^2\\ 0, & \text{otherwise}\end{cases}.$$

(2) The region D over which to integrate is shaded in Figure 3.45.

$$P(0\leqslant X\leqslant 1/2)=\iint_D f(x,y)\,\mathrm{d}x\mathrm{d}y=\int_0^{\frac{1}{2}}\left(\int_0^{1-x^2}\frac{5}{4}(x^2+y)\,\mathrm{d}y\right)\mathrm{d}x=\frac{79}{256}.$$

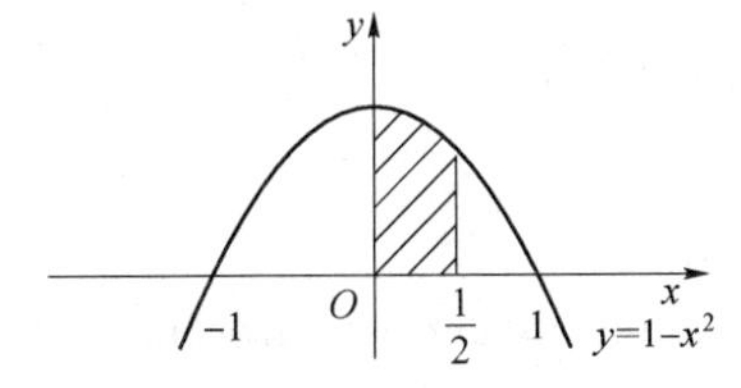

Figure 3.45

(3) The region D over which to integrate is shaded in Figure 3.46.

$$\begin{aligned}&P(Y\leqslant X+1)\\&=\iint_D f(x,y)\,\mathrm{d}x\mathrm{d}y\\&=\int_{-1}^{0}\left(\int_0^{x+1}\frac{5}{4}(x^2+y)\,\mathrm{d}y\right)\mathrm{d}x+\int_0^1\left(\int_0^{1-x^2}\frac{5}{4}(x^2+y)\,\mathrm{d}y\right)\mathrm{d}x\\&=\frac{13}{16}.\end{aligned}$$

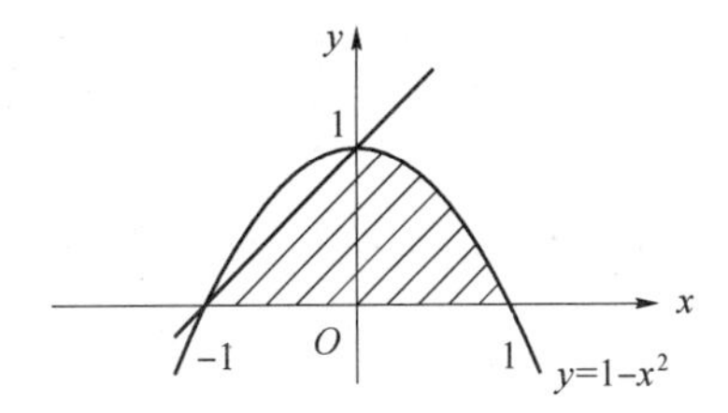

Figure 3.46

(4) $P(Y=X^2)=0$.

Problem 3.32 Suppose that X and Y are random variables such that (X,Y) must belong to the rectangle in the xy-plane containing all points (x,y) for which $0\leqslant x\leqslant 3$ and $0\leqslant y\leqslant 4$. Suppose also that the joint CDF of X and Y at every point (x,y) in this rectangle is specified as follows:

$$F(x,y)=\frac{1}{156}xy(x^2+y).$$

Determine:

(1) $P(1\leqslant X\leqslant 2 \text{ and } 1\leqslant Y\leqslant 2)$;

(2) $P(2\leqslant X\leqslant 4 \text{ and } 2\leqslant Y\leqslant 4)$;

(3) the CDF of Y;

(4) the joint PDF of X and Y;

(5) $P(Y\leqslant X)$.

Solution:

(1) Since the joint CDF is continuous and is twice-differentiable in the given rectangle, the joint distribution of X and Y is continuous. Therefore,

$$\begin{aligned}P(1\leqslant X\leqslant 2 \text{ and } 1\leqslant Y\leqslant 2) &= F(2,2)-F(1,2)-F(2,1)+F(1,1)\\ &= \frac{24}{156}-\frac{6}{156}-\frac{10}{156}+\frac{2}{156}\\ &= \frac{5}{78}.\end{aligned}$$

(2)

$$\begin{aligned}P(2\leqslant X\leqslant 4 \text{ and } 2\leqslant Y\leqslant 4) &= P(2\leqslant X\leqslant 3 \text{ and } 2\leqslant Y\leqslant 4)\\ &= F(3,4)-F(2,4)-F(3,2)+F(2,2)\\ &= 1-\frac{64}{156}-\frac{66}{156}+\frac{24}{156}\\ &= \frac{25}{78}.\end{aligned}$$

(3) Since Y must lie in the interval $0\leqslant y\leqslant 4$, $F_Y(y)=0$ for $y<0$ and $F_Y(y)=1$ for $y>4$. For $0\leqslant y\leqslant 4$,

$$F_Y(y)=\lim_{x\to+\infty}F(x,y)=\lim_{x\to 3}\frac{1}{156}xy(x^2+y)=\frac{1}{52}y(9+y).$$

The CDF of Y is

$$F_Y(y)=\begin{cases}0, & y<0\\ \dfrac{1}{52}y(9+y), & 0\leqslant y\leqslant 4\\ 1, & y>4\end{cases}.$$

(4) The PDF of (X,Y) is

$$f(x,y)=\frac{\partial^2 F(x,y)}{\partial x\partial y}=\begin{cases}\dfrac{1}{156}(3x^2+2y), & 0\leqslant x\leqslant 3, 0\leqslant y\leqslant 4\\ 0, & \text{otherwise}\end{cases}.$$

(5) The region D over which to integrate is shaded in Figure 3.47.

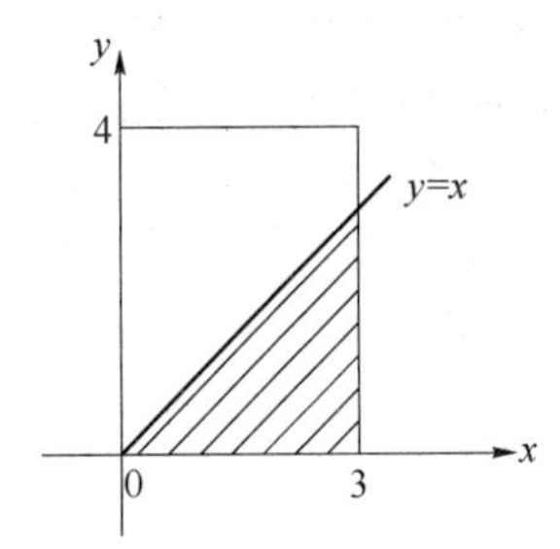

Figure 3.47

$$P(Y \leqslant X) = \iint_D f(x,y)\,dx\,dy = \int_0^3 \left(\int_0^x \frac{1}{156}(3x^2 + 2y)\,dy \right) dx = \frac{93}{208}.$$

Problem 3.33 Suppose that X and Y have a discrete joint distribution for which the joint PF is defined as follows:

$$f(x,y) = \begin{cases} \frac{1}{30}(x+y), & x=0,1,2 \text{ and } y=0,1,2,3 \\ 0, & \text{otherwise} \end{cases}.$$

Are X and Y independent?

Solution:

We first determine the marginal PFs of X and Y. For $x=0,1,2$, we have

$$f_X(x) = \sum_{y=0}^{3} f(x,y) = \frac{1}{30}(4x + 6) = \frac{1}{15}(2x + 3).$$

Similarly, for $y=0,1,2,3$, we have

$$f_Y(y) = \sum_{x=0}^{2} f(x,y) = \frac{1}{30}(3 + 3y) = \frac{1}{10}(1 + y).$$

X and Y are not independent because it is not true that $f(x,y)=f_X(x)f_Y(y)$ for all possible values of x and y.

Problem 3.34 Suppose that the joint PDF of X and Y is as follows:

$$f(x,y) = \begin{cases} \frac{15}{4}x^2, & 0 \leqslant y \leqslant 1-x^2 \\ 0, & \text{otherwise} \end{cases}.$$

Are X and Y independent?

Solution:

The support (the region where $f(x,y)$ is non-zero) is plotted in Figure 3.48.

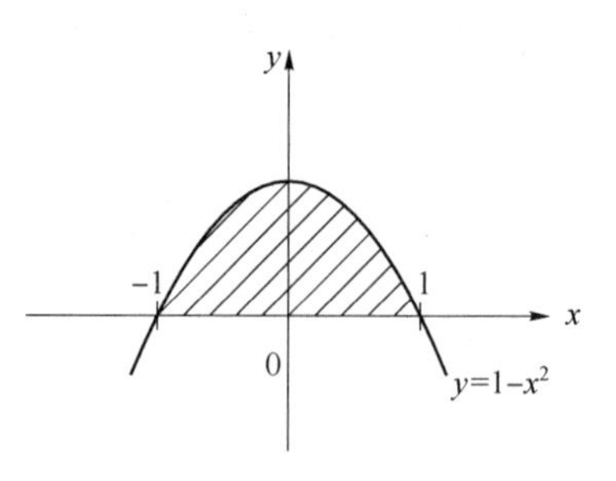

Figure 3.48

Find the marginal PDF $f_X(x)$ of X. It can be seen that the possible values of X are confined to the interval $-1\leqslant x\leqslant 1$. Hence, $f_X(x)=0$ for values of x outside this interval. For $-1\leqslant x\leqslant 1$, we have

$$\begin{aligned} f_X(x) &= \int_{-\infty}^{+\infty} f(x,y)\,\mathrm{d}y \\ &= \int_0^{1-x^2} \frac{15}{4}x^2\,\mathrm{d}y \\ &= \frac{15}{4}x^2(1-x^2). \end{aligned}$$

The PDF of X is

$$f_X(x)=\begin{cases} \frac{15}{4}x^2(1-x^2), & -1\leqslant x\leqslant 1 \\ 0, & \text{otherwise} \end{cases}.$$

Find the marginal PDF $f_Y(y)$ of Y. Similarly, it can be seen that the possible values of Y are confined to the interval $0\leqslant y\leqslant 1$. Hence, $f_Y(y)=0$ for values of y outside this interval. For $0\leqslant y\leqslant 1$, we have

$$f_Y(y)=\int_{-\infty}^{+\infty} f(x,y)\,\mathrm{d}x = \int_{-\sqrt{1-y}}^{\sqrt{1-y}} \frac{15}{4}x^2\,\mathrm{d}x = \frac{5}{2}(1-y)^{\frac{3}{2}}.$$

The PDF of Y is

$$f_Y(y)=\begin{cases} \frac{5}{2}(1-y)^{\frac{3}{2}}, & 0\leqslant y\leqslant 1 \\ 0, & \text{otherwise} \end{cases}.$$

In the support region, $f(x,y)\neq f_X(x)f_Y(y)$. Therefore, X and Y are not independent.

Problem 3.35 Suppose that the joint PDF of X and Y is as follows:

$$f(x,y)=\begin{cases} 2x\,\mathrm{e}^{-y}, & 0\leqslant x\leqslant 1 \text{ and } 0<y<+\infty \\ 0, & \text{otherwise} \end{cases}.$$

Are X and Y independent?

Solution:

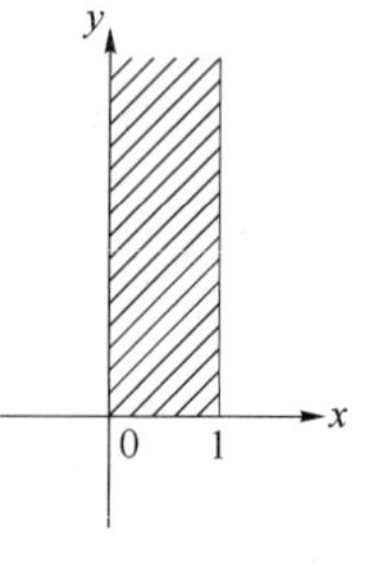

Figure 3.49

The support (the region where $f(x,y)$ is non-zero) is plotted in Figure 3.49.

Find the marginal PDF $f_X(x)$ of X. It can be seen that the possible values of X are confined to the interval $0\leqslant x\leqslant 1$. Hence, $f_X(x)=0$ for values of x outside this interval. For $0\leqslant x\leqslant 1$, we have

$$f_X(x)=\int_{-\infty}^{+\infty} f(x,y)\,\mathrm{d}y = \int_0^{+\infty} 2x\,\mathrm{e}^{-y}\,\mathrm{d}y = 2x.$$

The PDF of X is

$$f_X(x)=\begin{cases}2x, & 0\leqslant x\leqslant 1\\ 0, & \text{otherwise}\end{cases}.$$

Find the marginal PDF $f_Y(y)$ of Y. Similarly, it can be seen that the possible values of Y are confined to the interval $0<y<+\infty$. Hence, $f_Y(y)=0$ for values of y outside this interval. For $0<y<+\infty$, we have

$$f_Y(y)=\int_{-\infty}^{+\infty} f(x,y)\,dx=\int_0^1 2x\,e^{-y}dx=e^{-y}.$$

The PDF of Y is

$$f_Y(y)=\begin{cases}e^{-y}, & 0<y<+\infty\\ 0, & \text{otherwise}\end{cases}.$$

Therefore, $f(x,y)=f_X(x)f_Y(y)$, X and Y are independent.

Problem 3.36 Suppose that a point (X,Y) is chosen at random from the disc S defined as follows: $S=\{(x,y)\mid x^2+y^2\leqslant 1\}$.

(1) Determine the joint PDF of X and Y, the marginal PDF of X, and the marginal PDF of Y.

(2) Are X and Y independent?

 **Solution:**

(1) The joint PDF of (X,Y) is

$$f(x,y)=\begin{cases}\dfrac{1}{\pi}, & x^2+y^2\leqslant 1\\ 0, & \text{otherwise}\end{cases}.$$

The support region is plotted in Figure 3.50. Find the marginal PDF of Y.

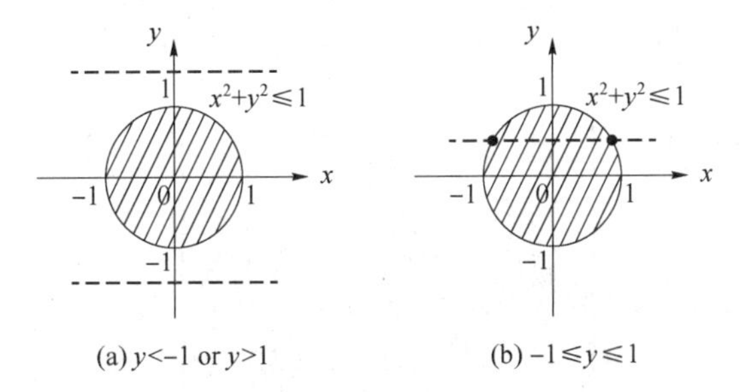

(a) $y<-1$ or $y>1$ (b) $-1\leqslant y\leqslant 1$

Figure 3.50

- If $y<-1$ or $y>1$, $f_Y(y)=0$.
- If $-1\leqslant y\leqslant 1$,

$$f_Y(y)=\int_{-\infty}^{+\infty} f(x,y)\,dx$$

$$= \int_{-\sqrt{1-y^2}}^{\sqrt{1-y^2}} \frac{1}{\pi} dx$$

$$= \frac{2\sqrt{1-y^2}}{\pi}.$$

The marginal PDF of Y is

$$f_Y(y) = \begin{cases} \frac{2\sqrt{1-y^2}}{\pi}, & -1 \leqslant y \leqslant 1 \\ 0, & \text{otherwise} \end{cases}.$$

According to the symmetry, the marginal PDF of X is

$$f_X(x) = \begin{cases} \frac{2\sqrt{1-x^2}}{\pi}, & -1 \leqslant x \leqslant 1 \\ 0, & \text{otherwise} \end{cases}.$$

(2) In the support region, $f(x,y) \neq f_X(x) f_Y(y)$. Therefore, X and Y are not independent.

Problem 3.37 Suppose that a point (X, Y) is chosen at random from the disk S defined as follows:

$$S = \{(x,y) \mid (x-1)^2 + (y+2)^2 \leqslant 9\}.$$

Determine:

(1) the conditional PDF of Y for every given value of X;

(2) $P(Y>0 \mid X=2)$.

 **Solution**:

(1) The joint PDF of (X, Y) is

$$f(x,y) = \begin{cases} \frac{1}{9\pi}, & (x-1)^2 + (y+2)^2 \leqslant 9 \\ 0, & \text{otherwise} \end{cases}.$$

The support region is plotted in Figure 3.51. Find the marginal PDF of X.

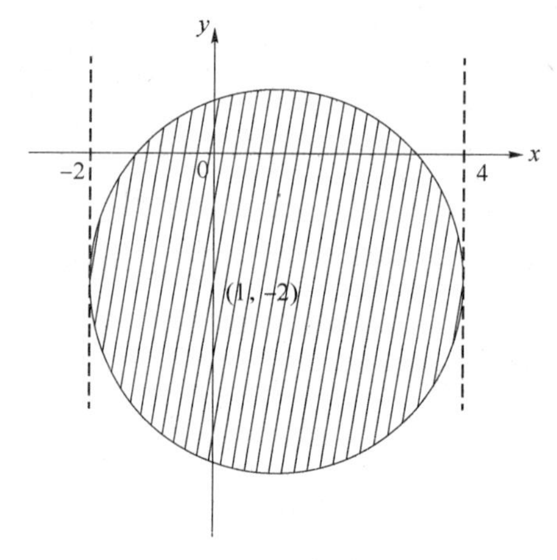

Figure 3.51

- If $x<-2$ or $x>4$, $f_X(x)=0$.
- If $-2\leqslant x\leqslant 4$,

$$f_X(x)=\int_{-\infty}^{+\infty} f(x,y)\,\mathrm{d}y$$

$$=\int_{-2-\sqrt{9-(x-1)^2}}^{-2+\sqrt{9-(x-1)^2}} \frac{1}{9\pi}\mathrm{d}y$$

$$=\frac{2\sqrt{9-(x-1)^2}}{9\pi}.$$

The marginal PDF of X is

$$f_X(x)=\begin{cases}\dfrac{2\sqrt{9-(x-1)^2}}{9\pi}, & -2\leqslant x\leqslant 4\\ 0, & \text{otherwise}\end{cases}.$$

The conditional PDF of Y for every given X is

$$f_{Y|X}(y|x)=\frac{f(x,y)}{f_X(x)}=\begin{cases}\dfrac{1}{2\sqrt{9-(x-1)^2}}, & (x-1)^2+(y+2)^2\leqslant 9\\ 0, & \text{otherwise}\end{cases}.$$

(2) The conditional PDF of Y for given $X=2$ is

$$f_{Y|X}(y|x=2)=\begin{cases}\dfrac{\sqrt{2}}{8}, & -2-2\sqrt{2}\leqslant y\leqslant -2+2\sqrt{2}\\ 0, & \text{otherwise}\end{cases}.$$

$$P(Y>0|X=2)=\int_0^{+\infty} f_{Y|X}(y|x=2)\,\mathrm{d}y=\int_0^{-2+2\sqrt{2}}\frac{\sqrt{2}}{8}\mathrm{d}y=\frac{2-\sqrt{2}}{4}.$$

Problem 3.38 Suppose that the joint PDF of two random variables X and Y is as follows:

$$f(x,y)=\begin{cases}\dfrac{3}{16}(4-2x-y), & x>0, y>0 \text{ and } 2x+y<4\\ 0, & \text{otherwise}\end{cases}.$$

Determine:

(1) the conditional PDF of Y for every given value of X;

(2) $P\left(Y\geqslant 2 \middle| X=\dfrac{1}{2}\right)$.

 **Solution**:

(1) The support region is plotted in Figure 3.52. Find the marginal PDF of X

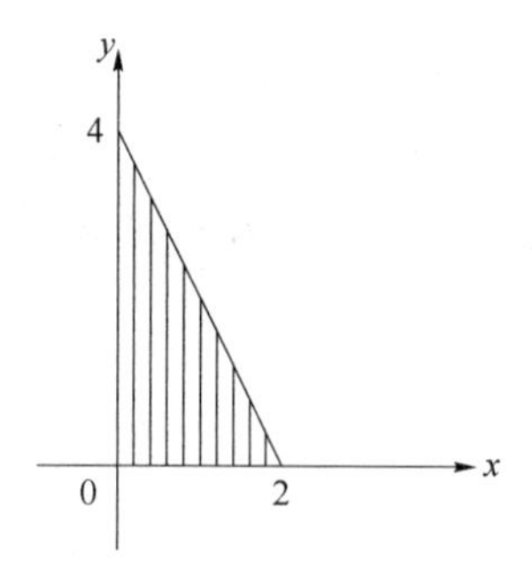

Figure 3.52

- If $x \leqslant 0$ or $x \geqslant 2, f_X(x) = 0$.
- If $0<x<2$,

$$\begin{aligned} f_X(x) &= \int_{-\infty}^{+\infty} f(x,y)\,\mathrm{d}y \\ &= \int_0^{4-2x} \frac{3}{16}(4-2x-y)\,\mathrm{d}y \\ &= \frac{3}{8}(x-2)^2. \end{aligned}$$

The marginal PDF of X is

$$f_X(x) = \begin{cases} \frac{3}{8}(x-2)^2, & 0<x<2 \\ 0, & \text{otherwise} \end{cases}.$$

The conditional PDF of Y for every given value of X is

$$f_{Y|X}(y \mid x) = \frac{f(x,y)}{f_X(x)} = \begin{cases} \frac{4-2x-y}{2(x-2)^2}, & x>0, y>0 \text{ and } 2x+y<4 \\ 0, & \text{otherwise} \end{cases}.$$

(2) The conditional PDF of Y for given $X = \frac{1}{2}$ is

$$f_{Y|X}\left(y \mid x = \frac{1}{2}\right) = \begin{cases} \frac{2(3-y)}{9}, & 0<y<3 \\ 0, & \text{otherwise} \end{cases}.$$

$$P\left(Y \geqslant 2 \mid X = \frac{1}{2}\right) = \int_2^{+\infty} f_{Y|X}\left(y \mid x = \frac{1}{2}\right)\mathrm{d}y = \int_2^3 \frac{2(3-y)}{9}\mathrm{d}y = \frac{1}{9}.$$

Problem 3.39 Suppose that $X_1, \cdots, X_n$ form a random sample of size n from the uniform distribution on the interval $[0,1]$ and that $Y_n = \max\{X_1, \cdots, X_n\}$. Find the smallest value of n such that $P(Y_n \geqslant 0.99) \geqslant 0.95$.

Solution:

$$\begin{aligned} P(Y_n \geqslant 0.99) &= 1 - P(Y_n < 0.99) \\ &= 1 - P(\max\{X_1, \cdots, X_n\} < 0.99) \end{aligned}$$

$$= 1-P(X_1<0.99,\cdots,X_n<0.99)$$
$$= 1-P(X_1<0.99)\times\cdots\times P(X_n<0.99)$$
$$= 1-0.99^n.$$

Thus, $1-0.99^n \geqslant 0.95$, $n \geqslant \dfrac{\ln 0.05}{\ln 0.99} \approx 298.07$, the smallest value of n is 299.

Problem 3.40 Suppose that the random variables X and Y have the following joint PDF:

$$f(x,y)=\begin{cases} 8xy, & 0\leqslant x\leqslant y\leqslant 1 \\ 0, & \text{otherwise} \end{cases}.$$

Also, let $U=\dfrac{X}{Y}$ and $V=Y$.

(1) Determine the joint PDF of U and V.

(2) Are X and Y independent?

(3) Are U and V independent?

Solution:

(1) The joint PDF of (X,Y) is

$$f(x,y)=\begin{cases} 8xy, & 0\leqslant x\leqslant y\leqslant 1 \\ 0, & \text{otherwise} \end{cases}.$$

$$\begin{cases} U=r_1(X,Y)=\dfrac{X}{Y} \\ V=r_2(X,Y)=Y \end{cases}$$

$$S=\{(x,y)\mid 0\leqslant x\leqslant y\leqslant 1\}$$

r_1, r_2 define a one-to-one differentiable transformation of S onto a subset T of $\mathbb{R}^2$, as shown in Figure 3.53. The inverse functions can be found as

$$\begin{cases} X=s_1(U,V)=UV \\ Y=s_2(U,V)=V \end{cases}$$

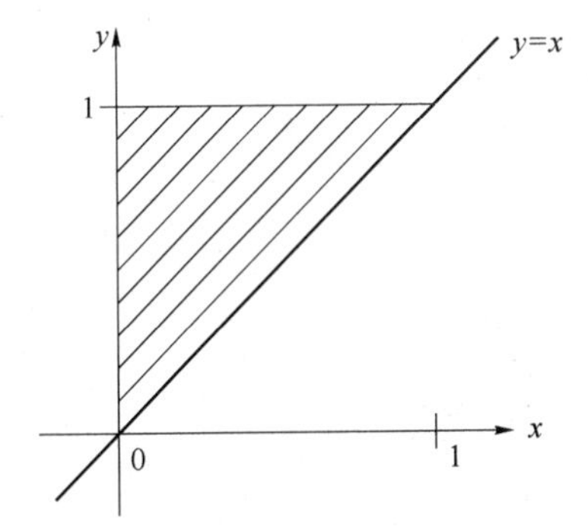

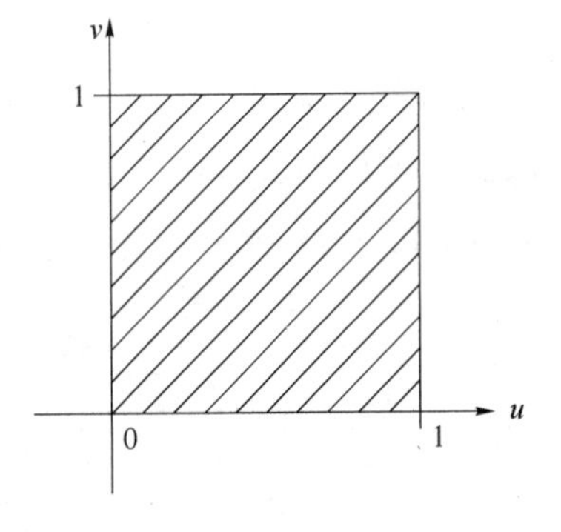

Figure 3.53

Therefore, $0\leqslant UV\leqslant V\leqslant 1$, and $T=\{(u,v)\mid 0\leqslant u\leqslant 1, 0\leqslant v\leqslant 1\}$.

$$J=\begin{vmatrix}\frac{\partial s_1}{\partial u} & \frac{\partial s_1}{\partial v}\\ \frac{\partial s_2}{\partial u} & \frac{\partial s_2}{\partial v}\end{vmatrix}=\begin{vmatrix}v & u\\ 0 & 1\end{vmatrix}=v,\ |J|=v.$$

Therefore, the joint PDF of (U,V) is

$$g(u,v)=\begin{cases}f(s_1,s_2)\times|J|, & (u,v)\in T\\ 0, & \text{otherwise}\end{cases}$$
$$=\begin{cases}8\times(uv)\times v\times v, & (u,v)\in T\\ 0, & \text{otherwise}\end{cases}$$
$$=\begin{cases}8u\,v^3, & 0\leqslant u\leqslant 1, 0\leqslant v\leqslant 1\\ 0, & \text{otherwise}\end{cases}.$$

(2) The joint PDF of (X,Y) is

$$f(x,y)=\begin{cases}8xy, & 0\leqslant x\leqslant y\leqslant 1\\ 0, & \text{otherwise}\end{cases}.$$

Its support region is $S=\{(x,y)\mid 0\leqslant x\leqslant y\leqslant 1\}$.

①Find the marginal PDF $f_X(x)$ of X. The support region is plotted in Figure 3.54.

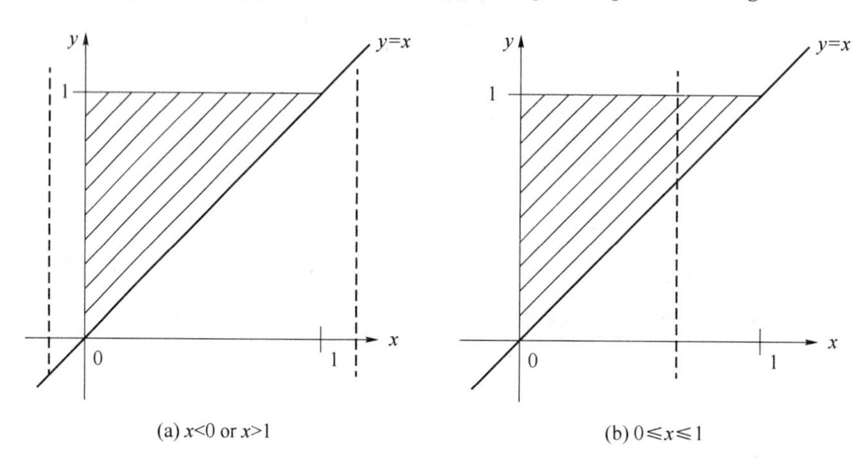

Figure 3.54

- If $x<0$ or $x>1$, $f_X(x)=0$.
- If $0\leqslant x\leqslant 1$,

$$f_X(x)=\int_{-\infty}^{+\infty}f(x,y)\,\mathrm{d}y$$
$$=\int_x^1 8xy\mathrm{d}y$$
$$=4xy^2\Big|_{y=x}^{y=1}$$
$$=4x(1-x^2).$$

In summary,

$$f_X(x)=\begin{cases}4x(1-x^2), & 0\leqslant x\leqslant 1\\ 0, & \text{otherwise}\end{cases}.$$

②Find the marginal PDF $f_Y(y)$ of Y. The support region is plotted in Figure 3.55.

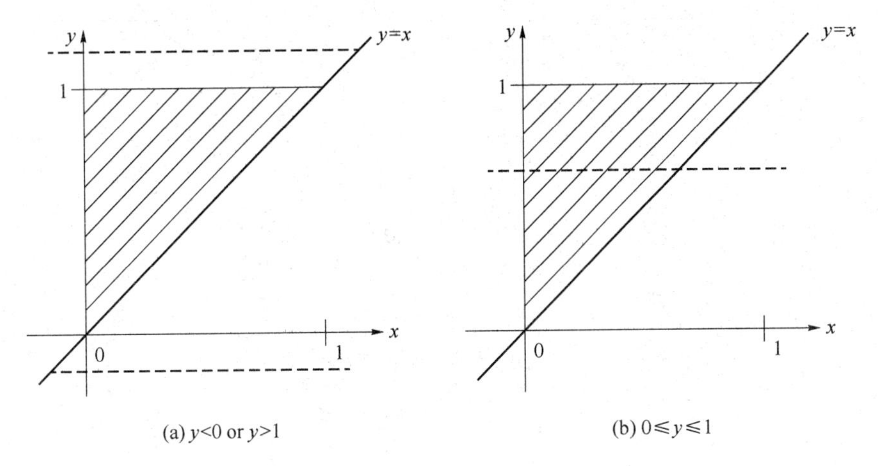

Figure 3.55

- If $y<0$ or $y>1$, $f_Y(y)=0$.
- If $0\leqslant y\leqslant 1$,

$$\begin{aligned}f_Y(y) &= \int_{-\infty}^{+\infty} f(x,y)\,\mathrm{d}x\\ &= \int_0^y 8xy\mathrm{d}x\\ &= 4yx^2 \,\Big|_{x=0}^{x=y}\\ &= 4\,y^3.\end{aligned}$$

In summary,

$$f_Y(y)=\begin{cases}4\,y^3, & 0\leqslant y\leqslant 1\\ 0, & \text{otherwise}\end{cases}.$$

Over the support set S, $8xy\neq 4x(1-x^2)\times 4\,y^3$, $f(x,y)\neq f_X(x)\times f_Y(y)$. Therefore, X and Y are not independent.

(3) The joint PDF of (U,V) is

$$g(u,v)=\begin{cases}8u\,v^3, & 0\leqslant u\leqslant 1, 0\leqslant v\leqslant 1\\ 0, & \text{otherwise}\end{cases}.$$

Its support region is $T=\{(u,v)\mid 0\leqslant u\leqslant 1, 0\leqslant v\leqslant 1\}$

①Find the marginal PDF $g_U(u)$ of U. The support region is plotted in Figure 3.56.

- If $u<0$ or $u>1$, $g_U(u)=0$.
- If $0\leqslant u\leqslant 1$,

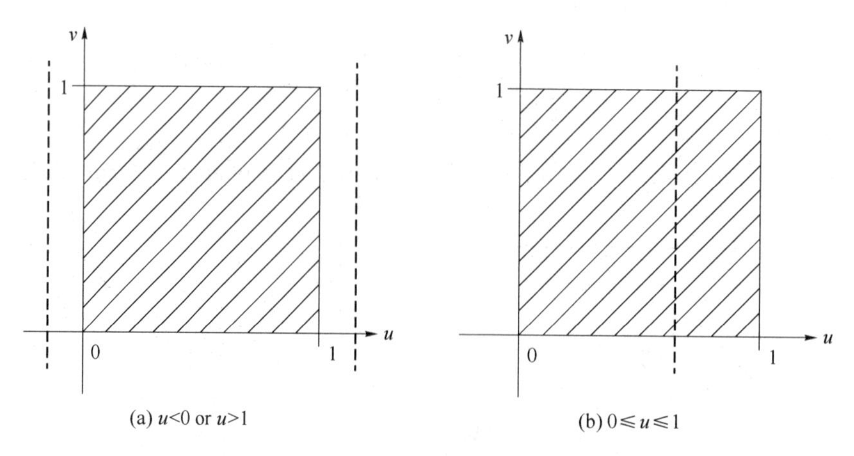

(a) $u<0$ or $u>1$

(b) $0\leqslant u\leqslant 1$

Figure 3.56

$$\begin{aligned} g_U(u) &= \int_{-\infty}^{+\infty} g(u,v)\,\mathrm{d}v \\ &= \int_0^1 8u\,v^3\,\mathrm{d}v \\ &= 2uv^4 \Big|_{v=0}^{v=1} \\ &= 2u. \end{aligned}$$

In summary,

$$g_U(u)=\begin{cases}2u, & 0\leqslant u\leqslant 1\\ 0, & \text{otherwise}\end{cases}.$$

②Find the marginal PDF $g_V(v)$ of V. The support region is plotted in Figure 3.57.

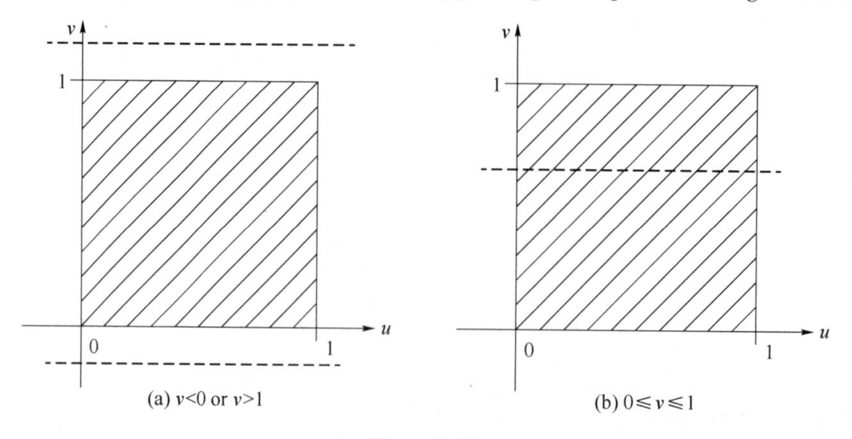

(a) $v<0$ or $v>1$

(b) $0\leqslant v\leqslant 1$

Figure 3.57

- If $v<0$ or $v>1$, $f_V(v)=0$.
- If $0\leqslant v\leqslant 1$,

$$\begin{aligned} g_V(v) &= \int_{-\infty}^{+\infty} g(u,v)\,\mathrm{d}u \\ &= \int_0^1 8uv^3\,\mathrm{d}u \end{aligned}$$

$$= 4v^3u^2 \mid_{u=0}^{u=1}$$
$$= 4v^3.$$

In summary,

$$g_V(v) = \begin{cases} 4v^3, & 0 \leqslant v \leqslant 1 \\ 0, & \text{otherwise} \end{cases}.$$

Since $g(u,v) = g_U(u) \times g_V(v)$ holds in the uv-plane, U and V are independent.

Chapter 4 Characteristics of Random Variables

Summary of Knowledge

Exercise Solutions

Problem 4.1 Suppose a discrete random variable X has the PF given by

X	3	5	7	8	9	10	12	16
$P(X=x)$	0.08	0.10	0.16	0.25	0.20	0.03	0.13	0.05

Find: (1) $E(X)$; (2) $V(X)$; (3) $E(X^2-4X)$.

Solution:

(1) $E(X)=3\times0.08+5\times0.10+7\times0.16+8\times0.25+9\times0.20+10\times0.03+12\times0.13+16\times0.05=8.32$.

(2) $E(X^2)=3^2\times0.08+5^2\times0.10+7^2\times0.16+8^2\times0.25+9^2\times0.20+10^2\times0.03+12^2\times0.13+16^2\times0.05=77.78$.

$V(X)=E(X^2)-[E(X)]^2=77.78-8.32^2=8.557\,6$.

(3) $E(X^2-4X)=E(X^2)-4E(X)=77.78-4\times8.32=44.5$.

Problem 4.2 Let X be a continuous random variable with the following PDF

$$f(x)=\begin{cases}x, & 0<x\leqslant 1\\ 2-x, & 1<x<2\\ 0, & \text{otherwise}\end{cases}.$$

Determine: (1) $E(X)$; (2) $V(X)$.

Solution:

$$E(X)=\int_{-\infty}^{+\infty}xf(x)\,\mathrm{d}x=\int_0^1 x^2\,\mathrm{d}x+\int_1^2 x(2-x)\,\mathrm{d}x=1.$$

$$E(X^2)=\int_{-\infty}^{+\infty}x^2f(x)\,dx=\int_0^1 x^3dx+\int_1^2 x^2(2-x)\,dx=\frac{7}{6}.$$

$$V(X)=E(X^2)-[E(X)]^2=\frac{1}{6}.$$

Problem 4.3 The PDF of X is $f(x)=\begin{cases}a+bx^2, & 0\leqslant x\leqslant 1\\ 0, & \text{else}\end{cases}$.

If $E(X)=\frac{3}{5}$, find a and b.

Solution:

By the property of PDF,

$$\int_{-\infty}^{+\infty}f(x)\,dx=\int_0^1(a+bx^2)\,dx=a+\frac{b}{3}=1.$$

On the other hand,

$$E(X)=\int_{-\infty}^{+\infty}xf(x)\,dx=\int_0^1 x(a+bx^2)\,dx=\frac{a}{2}+\frac{b}{4}=\frac{3}{5}.$$

Solving the two equations yields: $a=\frac{3}{5}, b=\frac{6}{5}$.

Problem 4.4 Let X be a continuous random variable with the following PDF

$$f(x)=\begin{cases}e^{-x}, & x>0\\ 0, & \text{otherwise}\end{cases}.$$

Let $Y=X+e^{-2X}$. Determine: (1) $E(Y)$; (2) $V(Y)$.

Solution:

$$E(Y)=E(X+e^{-2X})=\int_{-\infty}^{+\infty}(x+e^{-2x})f(x)\,dx=\int_0^{+\infty}(x+e^{-2x})\,e^{-x}dx=\frac{4}{3}.$$

$$E(Y^2)=E\,(X+e^{-2X})^2=\int_{-\infty}^{+\infty}(x+e^{-2x})^2f(x)\,dx=\int_0^{+\infty}(x+e^{-2x})^2\,e^{-x}dx=\frac{109}{45}.$$

$$V(Y)=E(Y^2)-[E(Y)]^2=\frac{29}{45}.$$

Problem 4.5 Let X and Y be two random variables having the following joint PF.

	$Y=1$	$Y=2$	$Y=3$
$X=-1$	0	$\frac{1}{15}$	$\frac{3}{15}$
$X=0$	$\frac{2}{15}$	$\frac{5}{15}$	$\frac{4}{15}$

Determine: (1) $E(X)$, $V(X)$; (2) $E(Y)$, $V(Y)$; (3) $E(XY)$, $V(XY)$.

Solution:

(1) The marginal PF of X is

	$X=-1$	$X=0$
P	$\frac{4}{15}$	$\frac{11}{15}$

$$E(X)=(-1)\times\frac{4}{15}+0\times\frac{11}{15}=-\frac{4}{15}.$$

$$E(X^2)=(-1)^2\times\frac{4}{15}+0^2\times\frac{11}{15}=\frac{4}{15}.$$

$$V(X)=E(X^2)-[E(X)]^2=\frac{44}{225}.$$

(2) The marginal PF of Y is

	$Y=1$	$Y=2$	$Y=3$
P	$\frac{2}{15}$	$\frac{6}{15}$	$\frac{7}{15}$

$$E(Y)=1\times\frac{2}{15}+2\times\frac{6}{15}+3\times\frac{7}{15}=\frac{7}{3}.$$

$$E(Y^2)=1^2\times\frac{2}{15}+2^2\times\frac{6}{15}+3^2\times\frac{7}{15}=\frac{89}{15}.$$

$$V(Y)=E(Y^2)-[E(Y)]^2=\frac{22}{45}.$$

(3) $E(XY)=(-1)\times1\times0+(-1)\times2\times\frac{1}{15}+(-1)\times3\times\frac{3}{15}+0\times1\times0+0\times2\times\frac{1}{15}+0\times3\times\frac{3}{15}=-\frac{11}{15}.$

$E[(XY)^2]=(-1)^2\times1^2\times0+(-1)^2\times2^2\times\frac{1}{15}+(-1)^2\times3^2\times\frac{3}{15}+0^2\times1^2\times0+0^2\times2^2\times\frac{1}{15}+0^2\times3^2\times\frac{3}{15}=\frac{31}{15}.$

$$V(XY)=E[(XY)^2]-[E(XY)]^2=\frac{344}{225}.$$

Problem 4.6 Suppose that X and Y have a continuous joint distribution for which the joint PDF is as follows:

$$f(x,y)=\begin{cases}12y^2, & 0\leqslant y\leqslant x\leqslant 1\\ 0, & \text{otherwise}\end{cases}.$$

Find $E(XY)$ and $V(XY)$.

Solution 1:

The support region is plotted in Figure 4.1.

$$E(XY)=\int_{-\infty}^{+\infty}\int_{-\infty}^{+\infty}xyf(x,y)\,\mathrm{d}x\mathrm{d}y=\int_0^1\left[\int_0^x(xy\times 12y^2)\,\mathrm{d}y\right]\mathrm{d}x=\frac{1}{2}.$$

$$E[(XY)^2]=\int_{-\infty}^{+\infty}\int_{-\infty}^{+\infty}(xy)^2f(x,y)\,\mathrm{d}x\mathrm{d}y$$

$$=\int_0^1\left[\int_0^x(xy)^2\times 12y^2\mathrm{d}y\right]\mathrm{d}x=\frac{3}{10}.$$

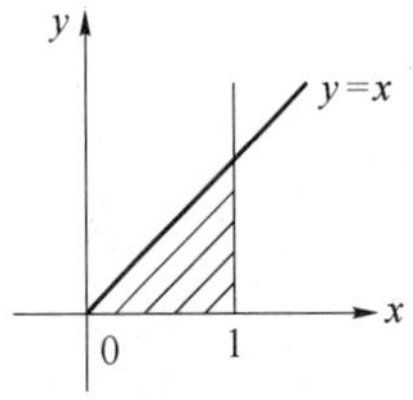

Figure 4.1

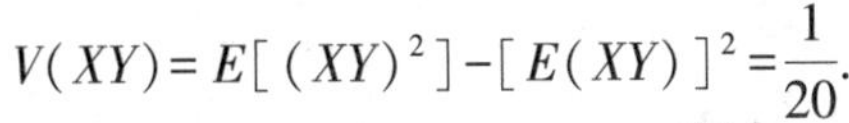

$$V(XY)=E[(XY)^2]-[E(XY)]^2=\frac{1}{20}.$$

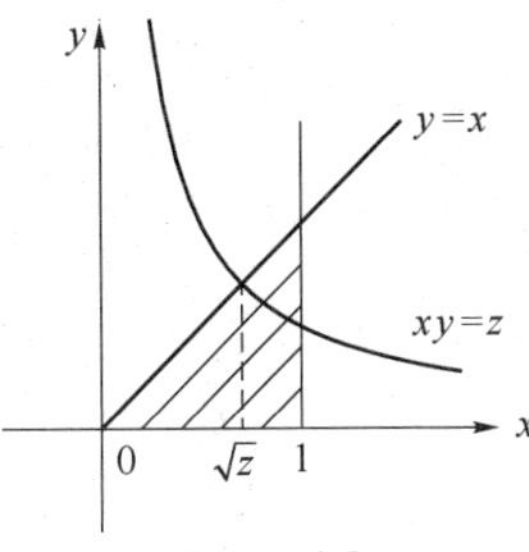

Figure 4.2

Solution 2:

Find the CDF $F_Z(z)$ of $Z=XY$. The support region is plotted in Figure 4.2.

- If $z\leqslant 0, F_Z(z)=P(\varnothing)=0$.
- If $0<z<1$,

$$F_Z(z)=P(XY\leqslant z)$$

$$=\iint_{xy\leqslant z}f(x,y)\,\mathrm{d}x\mathrm{d}y$$

$$=\int_0^{\sqrt{z}}\left(\int_0^x 12y^2\mathrm{d}y\right)\mathrm{d}x+\int_{\sqrt{z}}^1\left(\int_0^{\frac{z}{x}}12y^2\mathrm{d}y\right)\mathrm{d}x$$

$$=3z^2-2z^3.$$

- If $z\geqslant 1, F_Z(z)=1$.

The CDF of $Z=XY$ is

$$F_Z(z)=\begin{cases}0, & z\leqslant 0\\ 3z^2-2z^3, & 0<z<1.\\ 1, & z\geqslant 1\end{cases}$$

The PDF of $Z=XY$ is

$$f_Z(z)=F'_Z(z)=\begin{cases}6z-6z^2, & 0<z<1\\ 0, & \text{otherwise}\end{cases}.$$

$$E(XY)=E(Z)=\int_{-\infty}^{+\infty}zf_Z(z)\,\mathrm{d}x=\int_0^1 z(6z-6z^2)\,\mathrm{d}z=\frac{1}{2}.$$

$$E[(XY)^2]=E(Z^2)=\int_{-\infty}^{+\infty}z^2f_Z(z)\,\mathrm{d}x=\int_0^1 z^2(6z-6z^2)\,\mathrm{d}z=\frac{3}{10}.$$

$$V(XY)=V(Z)=E(Z^2)-[E(Z)]^2=\frac{1}{20}.$$

Problem 4.7 Suppose that X and Y have a continuous joint distribution for which the joint PDF is as follows:

$$f(x,y)=\begin{cases}3x, & 0<x<1, 0<y<x\\ 0, & \text{otherwise}\end{cases}.$$

Determine: (1) $E(X), V(X)$; (2) $E(Y), V(Y)$.

Solution:

The support region $\{(x,y) \mid 0<x<1, 0<y<x\}$ is plotted in Figure 4.3.

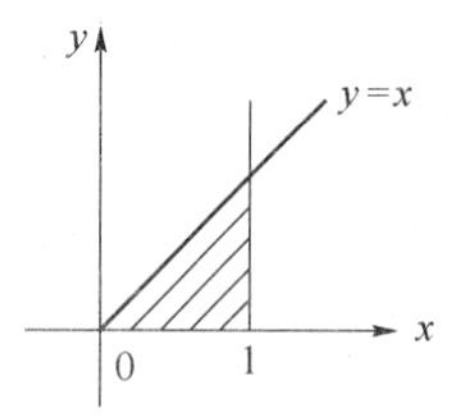

Figure 4.3

(1) $E(X)=\int_{-\infty}^{+\infty}\int_{-\infty}^{+\infty} xf(x,y)\,\mathrm{d}x\mathrm{d}y=\int_0^1\left(\int_0^x 3x^2\mathrm{d}y\right)\mathrm{d}x=\int_0^1 3x^3\mathrm{d}x=\frac{3}{4}.$

$E(X^2)=\int_{-\infty}^{+\infty}\int_{-\infty}^{+\infty} x^2f(x,y)\,\mathrm{d}x\mathrm{d}y=\int_0^1\left(\int_0^x 3x^3\mathrm{d}y\right)\mathrm{d}x=\int_0^1 3x^4\mathrm{d}x=\frac{3}{5}.$

$V(X)=E(X^2)-[E(X)]^2=\frac{3}{80}.$

(2) $E(Y)=\int_{-\infty}^{+\infty}\int_{-\infty}^{+\infty} yf(x,y)\,\mathrm{d}x\mathrm{d}y=\int_0^1\left(\int_0^x 3xy\mathrm{d}y\right)\mathrm{d}x=\int_0^1 \frac{3}{2}x^3\mathrm{d}x=\frac{3}{8}.$

$E(Y^2)=\int_{-\infty}^{+\infty}\int_{-\infty}^{+\infty} y^2f(x,y)\,\mathrm{d}x\mathrm{d}y=\int_0^1\left(\int_0^x 3xy^2\mathrm{d}y\right)\mathrm{d}x=\int_0^1 x^4\mathrm{d}x=\frac{1}{5}.$

$V(Y)=E(Y^2)-[E(Y)]^2=\frac{19}{320}.$

Problem 4.8 Suppose that a point is chosen at random on a stick of unit length and that the stick is broken into two pieces at that point. Find the expected value of the length of the longer piece.

Solution 1:

Let X denote the point at which the stick is broken, then X has the uniform distribution on the interval $[0,1]$.

$$f_X(x)=\begin{cases}1, & 0\leqslant x\leqslant 1\\ 0, & \text{otherwise}\end{cases}.$$

Let Y denote the length of the longer piece, then $Y=\max\{X,1-X\}$. Y is a function of X. Therefore,

$$\begin{aligned}E(Y)&=\int_{-\infty}^{+\infty}\max\{x,1-x\}\times f_X(x)\,\mathrm{d}x\\&=\int_0^1\max\{x,1-x\}\times 1\mathrm{d}x\\&=\int_0^{\frac{1}{2}}(1-x)\,\mathrm{d}x+\int_{\frac{1}{2}}^1 x\mathrm{d}x\\&=\frac{3}{4}.\end{aligned}$$

Solution 2:

Let X denote the point at which the stick is broken, then X has the uniform distribution on the interval $[0,1]$.

$$f_X(x)=\begin{cases}1, & 0\leqslant x\leqslant 1\\ 0, & \text{otherwise}\end{cases}.$$

Let Y denote the length of the longer piece, then $Y=\max\{X,1-X\}$. Y is a function of X.

Since the support of X is $[0,1]$, the support of Y is $\left[\frac{1}{2},1\right]$. Find the CDF $G(y)=P(Y\leqslant y)$ of Y, as shown in Figure 4.4.

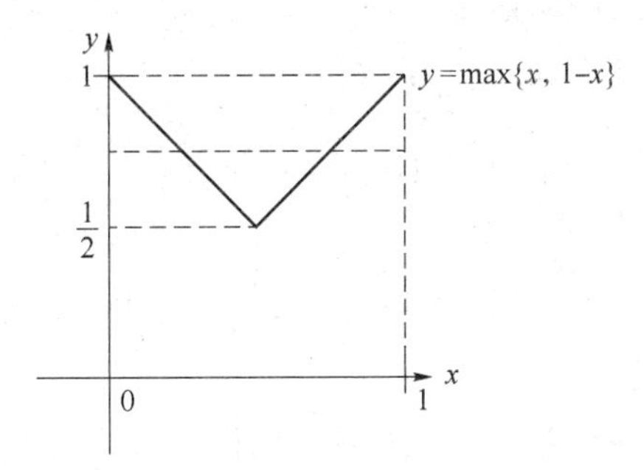

Figure 4.4

(1) If $y<\frac{1}{2}$, $G(y)=P(Y\leqslant y)=P(\varnothing)=0$.

(2) If $\frac{1}{2}\leqslant y\leqslant 1$.

$$\begin{aligned}G(y)&=P(Y\leqslant y)\\ &=P(1-y\leqslant X\leqslant y)\\ &=\int_{1-y}^{y}1\mathrm{d}x\\ &=2y-1.\end{aligned}$$

(3) If $1\leqslant y$, $G(y)=P(Y\leqslant y)=P(\Omega)=1$.

The CDF of Y is

$$G(y)=\begin{cases}0, & y<\frac{1}{2}\\ 2y-1, & \frac{1}{2}\leqslant y\leqslant 1\\ 1, & 1\leqslant y\end{cases}.$$

Therefore, The PDF of Y is

$$g(y)=G'(y)=\begin{cases}2, & \frac{1}{2}\leqslant y\leqslant 1\\ 0, & \text{otherwise}\end{cases}.$$

Therefore,

$$E(Y)=\int_{-\infty}^{+\infty} yg(y)\,\mathrm{d}y$$
$$=\int_{\frac{1}{2}}^{1} 2y\,\mathrm{d}y$$
$$=\frac{3}{4}.$$

Solution 3:

By the law of total probability,

$$G(y)=P(Y\leqslant y)$$
$$=P(\max\{X,1-X\}\leqslant y)$$
$$=P\left(\max\{X,1-X\}\leqslant y \mid X\leqslant\frac{1}{2}\right)P\left(X\leqslant\frac{1}{2}\right)+$$
$$P\left(\max\{X,1-X\}\leqslant y \mid X>\frac{1}{2}\right)P\left(X>\frac{1}{2}\right)$$
$$=P\left(1-X\leqslant y \mid X\leqslant\frac{1}{2}\right)P\left(X\leqslant\frac{1}{2}\right)+$$
$$P\left(X\leqslant y \mid X>\frac{1}{2}\right)P\left(X>\frac{1}{2}\right)$$
$$=\frac{1}{2}\left[P\left(1-y\leqslant X \mid X\leqslant\frac{1}{2}\right)+P\left(X\leqslant y \mid X>\frac{1}{2}\right)\right]$$

(1) If $y<\frac{1}{2}$, $G(y)=\frac{1}{2}[P(\varnothing)+P(\varnothing)]=0$.

(2) If $\frac{1}{2}\leqslant y\leqslant 1$,

$$G(y)=\frac{1}{2}\left[P\left(1-y\leqslant X \mid X\leqslant\frac{1}{2}\right)+P\left(X\leqslant y \mid X>\frac{1}{2}\right)\right]$$
$$=\frac{1}{2}\left[\frac{P\left(1-y\leqslant X\leqslant\frac{1}{2}\right)}{P\left(X\leqslant\frac{1}{2}\right)}+\frac{P\left(\frac{1}{2}<X\leqslant y\right)}{P\left(X>\frac{1}{2}\right)}\right]$$
$$=\frac{1}{2}-(1-y)+y-\frac{1}{2}$$
$$=2y-1.$$

(3) If $y\geqslant 1$, $G(y)=\frac{1}{2}[P(\Omega)+P(\Omega)]=1$.

The CDF of Y is

$$G(y)=\begin{cases}0, & y<\frac{1}{2}\\ 2y-1, & \frac{1}{2}\leqslant y\leqslant 1\\ 1, & y\geqslant 1\end{cases}.$$

The other steps are the same as solution 2.

Problem 4.9 Suppose that the random variables $X_1,\cdots,X_n$ form a random sample of size n from the uniform distribution on the interval $[0,1]$. Let $Y_1=\min\{X_1,\cdots,X_n\}$, and let $Y_n=\max\{X_1,\cdots,X_n\}$. Find $E(Y_1)$ and $E(Y_n)$.

 **Solution**:

The PDF and CDF of $X_i(i=1,\cdots,n)$ are given by

$$f(x)=\begin{cases}1, & 0\leqslant x\leqslant 1\\ 0, & \text{otherwise}\end{cases},\quad F(x)=\begin{cases}0, & x<0\\ x, & 0\leqslant x\leqslant 1\\ 1, & x>1\end{cases}.$$

$$F_{Y_n}(x)=P(Y_n\leqslant x)=P(X_1\leqslant x,\cdots,X_n\leqslant x)=[F(x)]^n=\begin{cases}0, & x<0\\ x^n, & 0\leqslant x\leqslant 1\\ 1, & x>1\end{cases}.$$

The PDF of Y_n is

$$f_{Y_n}(x)=n[F(x)]^{n-1}f(x)=\begin{cases}n x^{n-1}, & 0\leqslant x\leqslant 1\\ 0, & \text{otherwise}\end{cases}.$$

$$E(Y_n)=\int_{-\infty}^{+\infty}x f_{Y_n}(x)\,\mathrm{d}x=\int_0^1 n x^n\,\mathrm{d}x=\frac{n}{n+1}.$$

$$F_{Y_1}(x)=P(Y_1\leqslant x)=1-[1-F(x)]^n=\begin{cases}0, & x<0\\ 1-(1-x)^n, & 0\leqslant x\leqslant 1\\ 1, & x>1\end{cases}.$$

The PDF of Y_1 is

$$f_{Y_1}(x)=n[1-F(x)]^{n-1}f(x)=\begin{cases}n(1-x)^{n-1}, & 0\leqslant x\leqslant 1\\ 0, & \text{otherwise}\end{cases}.$$

$$E(Y_1)=\int_{-\infty}^{+\infty}x f_{Y_1}(x)\,\mathrm{d}x=\int_0^1 nx(1-x)^{n-1}\,\mathrm{d}x=\int_0^1(1-x)^n\,\mathrm{d}x=\frac{1}{n+1}.$$

Problem 4.10 Suppose that $X_1,\cdots,X_n$ form a random sample of size n from a continuous distribution with the following PDF:

$$f(x)=\begin{cases}2x, & 0<x<1\\ 0, & \text{otherwise}\end{cases}.$$

Let $Y_n=\max\{X_1,\cdots,X_n\}$. Evaluate $E(Y_n)$.

The CDF of $X_i(i=1,\cdots,n)$ is given by

$$F(x)=\int_{-\infty}^{x} f(t)\,\mathrm{d}t=\begin{cases}0, & x\leqslant 0\\ x^2, & 0<x<1.\\ 1, & x\geqslant 1\end{cases}$$

$$F_{Y_n}(x)=P(Y_n\leqslant x)=P(X_1\leqslant x,\cdots,X_n\leqslant x)=[F(x)]^n=\begin{cases}0, & x\leqslant 0\\ x^{2n}, & 0<x<1 .\\ 1, & x\geqslant 1\end{cases}$$

The PDF of Y_n is

$$f_{Y_n}(x)=n[F(x)]^{n-1}f(x)=\begin{cases}2n\,x^{2n-1}, & 0<x<1\\ 0, & \text{otherwise}\end{cases}.$$

$$E(Y_n)=\int_{-\infty}^{+\infty} x f_{Y_n}(x)\,\mathrm{d}x=\int_0^1 2n\,x^{2n}\,\mathrm{d}x=\frac{2n}{2n+1}.$$

Problem 4.11 Suppose that X and Y are IID and $X\sim B(1,0.4)$. Find $E[\max(X,Y)]$ and $E[\min(X,Y)]$.

All possible values of $\max(X,Y)$ are 0,1.

$$P(\max(X,Y)=0)=P(X=0,Y=0)=P(X=0)P(Y=0)=0.6\times 0.6=0.36.$$

$$P(\max(X,Y)=1)=1-0.36=0.64.$$

So the PF of $\max(X,Y)$ is given by

$\max(X,Y)$	0	1
P	0.36	0.64

$$E[\max(X,Y)]=0\times 0.36+1\times 0.64=0.64.$$

All possible values of $\min(X,Y)$ are 0,1.

$P(\min(X,Y)=1)=P(X=1,Y=1)=P(X=1)P(Y=1)=0.4\times 0.4=0.16.$

$P(\min(X,Y)=0)=1-0.16=0.84.$

So the PF of $\min(X,Y)$ is given by

$\min(X,Y)$	0	1
P	0.84	0.16

$$E[\min(X,Y)]=0\times 0.84+1\times 0.16=0.16.$$

Problem 4.12 A person has n keys and one lock. After every failed attempt to unlock, the person may pick a new key and discard the old one. What is the expected number of attempts needed to open the lock?

Solution 1:

Let X denote the number of attempts needed to open the lock.

$$P(X=1)=\frac{1}{n}$$

$$P(X=2)=\frac{n-1}{n}\times\frac{1}{n-1}=\frac{1}{n}$$

$$P(X=3)=\frac{n-1}{n}\times\frac{n-2}{n-1}\times\frac{1}{n-2}=\frac{1}{n}$$

$$\vdots$$

$$P(X=n)=\frac{n-1}{n}\times\cdots\times\frac{1}{2}\times 1=\frac{1}{n}$$

The PF of X is

X	1	2	3	$\cdots$	n
P	$\frac{1}{n}$	$\frac{1}{n}$	$\frac{1}{n}$	$\cdots$	$\frac{1}{n}$

Thus,

$$E(X)=\sum_{i=1}^{n} i\times P(X=i)=\frac{1}{n}\sum_{i=1}^{n} i=\frac{n+1}{2}.$$

Solution 2:

Let $X_i=\begin{cases} i, & \text{The } i\text{-th attempt opens the lock} \\ 0, & \text{otherwise} \end{cases}, i=1,\cdots,n.$

The number of attempts is $X=X_1+\cdots+X_n$. Here the $X_i's(i=1,2,\cdots,n)$ are not independent of each other and have different distributions.

$$P(X_1=1)=\frac{1}{n}$$

$$P(X_2=2)=\frac{n-1}{n}\times\frac{1}{n-1}=\frac{1}{n}$$

$$\vdots$$

$$P(X_i=i)=\frac{n-1}{n}\times\frac{n-2}{n-1}\times\cdots\times\frac{n-(i-1)}{n-(i-2)}\times\frac{1}{n-(i-1)}=\frac{1}{n}$$

$$\vdots$$

$$P(X_n=n)=\frac{n-1}{n}\times\cdots\times\frac{1}{2}\times 1=\frac{1}{n}$$

Therefore,

$$P(X_i=0)=1-\frac{1}{n}.$$

The PF of X_i is

X_i	0	i
P	$1-\frac{1}{n}$	$\frac{1}{n}$

$$E(X_i)=0\times P(X_i=0)+i\times P(X_i=i)=\frac{i}{n}.$$

According to the linear additivity of mathematical expectation,

$$E(X)=E(X_1+\cdots+X_n)=\frac{1}{n}+\frac{2}{n}+\cdots+\frac{n}{n}=\frac{n+1}{2}.$$

Solution 3:

Let Y be the number of unsuccessful unlocking attempts. The last attempt is successful. The total number of trials is $X=Y+1$. We can decompose Y as follows:

Let $X_i=\begin{cases}1, & \text{The first to the } i\text{-th attempt didn't unlock the lock} \\ 0, & \text{otherwise}\end{cases}, i=1,\cdots,n-1.$

$$P(X_i=1)=\frac{n-1}{n}\times\frac{n-2}{n-1}\times\cdots\times\frac{n-i}{n-i+1}=\frac{n-i}{n}, P(X_i=0)=\frac{i}{n}$$

The PF of X_i is

X_i	0	1
P	$\frac{i}{n}$	$\frac{n-i}{n}$

$$E(X_i)=0\times P(X_i=0)+1\times P(X_i=1)=\frac{n-i}{n}.$$

The unsuccessful unlocking times can be expressed as

$$Y=X_1+\cdots+X_{n-1}.$$

$$E(Y)=E(X_1+\cdots+X_{n-1})=\frac{n-1}{n}+\frac{n-2}{n}+\cdots+\frac{1}{n}=\frac{n-1}{2}.$$

$$E(X)=E(Y+1)=E(Y)+1=\frac{n-1}{2}+1=\frac{n+1}{2}.$$

Problem 4.13 A fair die is rolled 20 times. Find the expected sum of the 20 rolls.

Solution:

Let $X_i(i=1,2,\cdots,20)$ denote the number of the i-th roll, and let X be the sum of the 20 rolls. The PF of $X_i(i=1,2,\cdots,20)$ is

X_i	1	2	3	4	5	6
P	$\frac{1}{6}$	$\frac{1}{6}$	$\frac{1}{6}$	$\frac{1}{6}$	$\frac{1}{6}$	$\frac{1}{6}$

Then

$$E(X_i)=1\times\frac{1}{6}+2\times\frac{1}{6}+3\times\frac{1}{6}+4\times\frac{1}{6}+5\times\frac{1}{6}+6\times\frac{1}{6}=\frac{7}{2}.$$

$$E(X)=E\left(\sum_{i=1}^{20}X_i\right)=\sum_{i=1}^{20}E(X_i)=\frac{7}{2}\times 20=70.$$

Problem 4.14 Suppose that two persons A and B each randomly, and independently, choose 3 of 10 objects. Find the expected number of objects (1) chosen by both A and B; (2) not chosen by either A or B; (3) chosen by exactly one of A or B.

Solution 1:

(1) Let X be the number of objects that are selected by both A and B. To further simplify the problem we use indicator variables X_i. Let $X_i=1$ if object i is selected by both A and B, and $X_i=0$ otherwise, where $1\leqslant i\leqslant 10$. Then

$$E(X)=E\left(\sum_{i=1}^{n}X_i\right)=\sum_{i=1}^{10}E(X_i).$$

Now we must find $E(X_i)$. We know that X_i only takes on one of two values, $X_i=1$ or $X_i=0$. So, for the case of a sum of independent random indicator variables, $E(X_i)=1\times P(X_i=1)+0\times P(X_i=0)=P(X_i=1)$.

Each person can choose 3 of the 10 items. There are 3 ways to choose the item of interest, since a person can draw 3 objects. Since person A and B draw independently,

$$P(X_i=1)=\left(\frac{1\times C_9^2}{C_{10}^3}\right)^2=\left(\frac{3}{10}\right)^2.$$

Then,

$$E(X)=\sum_{i=1}^{10}E(X_i)=\sum_{i=1}^{10}\left(\frac{3}{10}\right)^2=10\times\left(\frac{3}{10}\right)^2=0.9.$$

(2) The principle is similar to part (1). Let Y be the number of objects that are not chosen by either A or B. Let $Y_i=1$ if object i is not chosen by A and is not chosen by B.

$$P(Y_i=1)=\left(\frac{C_9^3}{C_{10}^3}\right)^2=\left(\frac{7}{10}\right)^2.$$

Then,

$$E(Y)=\sum_{i=1}^{10}E(Y_i)=\sum_{i=1}^{10}\left(\frac{7}{10}\right)^2=10\times\left(\frac{7}{10}\right)^2=4.9.$$

(3) In this case, either person A draws object i and person B does not, or person B draws object i and person A does not. Let Z be the number of objects that are chosen by exactly one of A or B. Again, let $Z_i=1$ if exactly one of A or B draws object i, $Z_i=0$ otherwise. Thus,

$$P(Z_i=1)=2\times\left(\frac{C_9^3}{C_{10}^3}\right)\times\left(\frac{1\times C_9^2}{C_{10}^3}\right)=2\times\frac{7}{10}\times\frac{3}{10}.$$

Then,

$$E(Z)=\sum_{i=1}^{10}E(Z_i)=10\times\left(2\times\frac{7}{10}\times\frac{3}{10}\right)=4.2.$$

Solution 2:

(1) Let X be the number of objects that are selected by both A and B. X can take $\{0,1,2,3\}$.

$$P(X=0)=\frac{C_{10}^3\times C_7^3}{C_{10}^3\times C_{10}^3}=\frac{35}{120}$$

$$P(X=1)=\frac{C_{10}^3\times C_3^1\times C_7^2}{C_{10}^3\times C_{10}^3}=\frac{63}{120}$$

$$P(X=2)=\frac{C_{10}^3\times C_3^2\times C_7^1}{C_{10}^3\times C_{10}^3}=\frac{21}{120}$$

$$P(X=3)=\frac{C_{10}^3\times 1}{C_{10}^3\times C_{10}^3}=\frac{1}{120}$$

$$E(X)=0\times\frac{35}{120}+1\times\frac{63}{120}+2\times\frac{21}{120}+3\times\frac{1}{120}=0.9.$$

(2) Let Y be the number of objects that are not chosen by either A or B. Y can take $\{4,5,6,7\}$. $Y=10-(6-X)=4+X$.

$$E(Y)=E(4+X)=4+E(X)=4.9.$$

(3) Let Z be the number of objects that are chosen by exactly one of A or B. Z can take $\{0,2,4,6\}$. $Z+2X=6$.

$$E(Z)=E(6-2X)=6-2E(X)=4.2.$$

Problem 4.15 Six couples go to dinner together. The waiter seats the men randomly on one side of the table and the women randomly on the other side of the table. Find the expected value and variance of the number of couples who are seated across from each other.

Solution 1:

Let $X_i=\begin{cases}1, & \text{the } i\text{-th couple seated across from each other}\\ 0, & \text{otherwise}\end{cases}$ $(i=1,2,\cdots,6)$, then $X=\sum_{i=1}^{6}X_i$.

Here the $X_i's\,(i=1,2,\cdots,6)$ are not independent but have identical distributions.

$$P(X_i=1)=\frac{1}{6},P(X_i=0)=1-\frac{1}{6}=\frac{5}{6}.$$

Therefore, $E(X_i)=\frac{1}{6}, E(X)=\sum_{i=1}^{6}E(X_i)=1.$

$$
\begin{aligned}
E(X^2) &= E\left[\left(\sum_{i=1}^{6} X_i\right)^2\right] \\
&= E\left[\sum_{i=1}^{6} X_i^2 + \sum_{1\leqslant i\neq j\leqslant 6} X_i X_j\right] \\
&= \sum_{i=1}^{6} E[X_i^2] + \sum_{1\leqslant i\neq j\leqslant 6} E[X_i X_j] \\
&= \sum_{i=1}^{6}\left(1^2 \times \frac{1}{6}\right) + \sum_{1\leqslant i\neq j\leqslant 6} 1 \times 1 \times P(X_i = 1, X_j = 1) \\
&= 1 + \sum_{1\leqslant i\neq j\leqslant 6}\left(\frac{1}{6} \times \frac{1}{5}\right) \\
&= 2.
\end{aligned}
$$

Therefore, $V(X)=E(X^2)-[E(X)]^2=1$.

 **Solution 2:**

Let A_i be the event that the i-th couple seated across from each other ($i=1,\cdots,6$), and let X be the number of couples who are seated across from each other,

$$
\begin{aligned}
P(X=0) &= P(A_1^c \cap A_2^c \cap \cdots \cap A_6^c) \\
&= P\left[\left(\bigcup_{i=1}^{6} A_i\right)^c\right] \\
&= 1-P\left(\bigcup_{i=1}^{6} A_i\right) \\
&= 1-\left[1-\frac{1}{2!}+\frac{1}{3!}-\frac{1}{4!}+\cdots+(-1)^{6+1}\frac{1}{6!}\right] \\
&= \frac{1}{2!}-\frac{1}{3!}+\frac{1}{4!}-\frac{1}{5!}+\frac{1}{6!} \\
&= \frac{53}{144}.
\end{aligned}
$$

Similarly,

$$
\begin{aligned}
P(X=1) &= C_6^1\times\frac{1}{6}\times\left(\frac{1}{2!}-\frac{1}{3!}+\frac{1}{4!}-\frac{1}{5!}\right)=\frac{11}{30} \\
P(X=2) &= C_6^2\times\frac{1}{6}\times\frac{1}{5}\times\left(\frac{1}{2!}-\frac{1}{3!}+\frac{1}{4!}\right)=\frac{3}{16} \\
P(X=3) &= C_6^3\times\frac{1}{6}\times\frac{1}{5}\times\frac{1}{4}\times\left(\frac{1}{2!}-\frac{1}{3!}\right)=\frac{1}{18} \\
P(X=4) &= C_6^3\times\frac{1}{6}\times\frac{1}{5}\times\frac{1}{4}\times\frac{1}{3}\times\left(\frac{1}{2!}\right)=\frac{1}{48} \\
P(X=5) &= 0 \\
P(X=6) &= \frac{1}{6!}=\frac{1}{720}.
\end{aligned}
$$

The PF of X is

X	0	1	2	3	4	5	6
P	$\frac{53}{144}$	$\frac{11}{30}$	$\frac{3}{16}$	$\frac{1}{18}$	$\frac{1}{48}$	0	$\frac{1}{720}$

$$E(X)=0\times\frac{53}{144}+1\times\frac{11}{30}+2\times\frac{3}{16}+3\times\frac{1}{18}+4\times\frac{1}{48}+5\times0+6\times\frac{1}{720}=1.$$

$$E(X^2)=0^2\times\frac{53}{144}+1^2\times\frac{11}{30}+2^2\times\frac{3}{16}+3^2\times\frac{1}{18}+4^2\times\frac{1}{48}+5^2\times0+6^2\times\frac{1}{720}=2.$$

$$V(X)=E(X^2)-[E(X)]^2=1.$$

Problem 4.16 A total of n balls, numbered 1 through n, are put into n urns, also numbered 1 through n in such a way that ball i is equally likely to go into any of the urns $1,2,\cdots,i$. Find the expected number of empty urns.

Solution:

Let $X_i=\begin{cases}1, & \text{the } i\text{-th urn is empty}\\ 0, & \text{the } i\text{-th urn is not empty}\end{cases}$, then the number of empty urns $X=\sum\limits_{i=1}^{n}X_i$. Here $X_i(i=1,2,\cdots,n)$ satisfies

$$P(X_i=1)=\frac{i-1}{i}\times\frac{i}{i+1}\times\cdots\times\frac{n-1}{n}=\frac{i-1}{n}$$

$$P(X_i=0)=1-\frac{i-1}{n}$$

$$E(X_i)=0\times P(X_i=0)+1\times P(X_i=1)=\frac{i-1}{n}.$$

The expected number of empty urns is

$$E(X)=E\left(\sum_{i=1}^{n}X_i\right)=\sum_{i=1}^{n}E(X_i)=\sum_{i=1}^{n}\frac{i-1}{n}=\frac{n-1}{2}.$$

Problem 4.17 Suppose that three random variables X_1,X_2,X_3 form a random sample from the uniform distribution on $[0,1]$. Determine $E[(X_1-2X_2+X_3)^2]$.

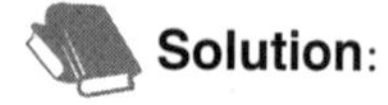

Solution:

$$\begin{aligned}&E[(X_1-2X_2+X_3)^2]\\&=E(X_1^2+4X_2^2+X_3^2-4X_1X_2+2X_1X_3-4X_2X_3)\\&=E(X_1^2)+4E(X_2^2)+E(X_3^2)-4E(X_1X_2)+2E(X_1X_3)-4E(X_2X_3).\end{aligned}$$

Since X_1,X_2, and X_3 are independent, $E(X_iX_j)=E(X_i)E(X_j)$ for $i\neq j$. Therefore, the above expectation can be written in the form:

$$E(X_1^2)+4E(X_2^2)+E(X_3^2)-4E(X_1)E(X_2)+2E(X_1)E(X_3)-4E(X_2)E(X_3).$$

Also, since each X_i has the uniform distribution on the interval $[0,1]$, then $E(X_i)=\frac{1}{2}$ and $E(X_i^2)=\int_0^1 x^2\mathrm{d}x=\frac{1}{3}$. Hence, the desired expectation has the value $\frac{1}{2}$.

Problem 4.18 Suppose that the random variable X has the uniform distribution on the interval $[0,1]$, that the random variable Y has the uniform distribution on the interval $[5,9]$, and that X and Y are independent. Suppose also that a rectangle is to be constructed for which the lengths of two adjacent sides are X and Y. Determine the expected value of the area of the rectangle.

 **Solution**:

The area of the rectangle is XY. Since X and Y are independent, $E(XY)=E(X)E(Y)$. Also, $E(X)=\frac{1}{2}$ and $E(Y)=7$. Therefore, $E(XY)=\frac{7}{2}$.

Problem 4.19 Suppose that a fair coin is tossed repeatedly until a head is obtained for the first time. (1) What is the expected number of tosses that will be required? (2) What is the expected number of tails that will be obtained before the first head is obtained?

 **Solution**:

(1) Let X denote the number of tosses that will be required.

$$P(X=k)=\left(1-\frac{1}{2}\right)^{k-1}\times\frac{1}{2}=\left(\frac{1}{2}\right)^k,k=1,2,3,\cdots.$$

X has the geometric distribution with parameter $p=\frac{1}{2}$. So the expected number of tosses is

$$E(X)=\sum_{k=1}^{+\infty}k\times\left(\frac{1}{2}\right)^k=2.$$

(2) Let Y be the number of tails that will be obtained, then $Y=X-1$. Therefore, the expected number of tails is $E(Y)=E(X-1)=E(X)-1=1$.

Problem 4.20 Suppose that X and Y are independent random variables for which $X\sim N(0,4)$, $Y\sim U(0,4)$. Find the values of $V(2X+3Y)$ and $V(2X-3Y)$.

 **Solution**:

$$V(2X+3Y)=4V(X)+9V(Y)=4\times4+9\times\frac{4^2}{12}=28.$$

$$V(2X-3Y)=4V(X)+9V(Y)=4\times4+9\times\frac{4^2}{12}=28.$$

Problem 4.21 Suppose $X \sim B(n,p)$. $E(X)=2.4, V(X)=1.44$. Find n and p.

Solution:

$E(X)=np=2.4$. $V(X)=np(1-p)=1.44$. Simultaneously solving the two equations, we can obtain $n=6, p=0.4$.

Problem 4.22 Suppose $X \sim U(a,b)$. $E(X)=3, V(X)=\frac{1}{3}$. Find $P(1<X<3)$.

Solution:

$E(X)=\frac{a+b}{2}=3, V(X)=\frac{(b-a)^2}{12}=\frac{1}{3}$, so that $a=2, b=4$. $X \sim U(2,4)$. The PDF of X is

$$f(x)=\begin{cases}\frac{1}{2}, & 2<x<4\\ 0, & \text{otherwise}\end{cases}.$$

Hence,

$$P(1<X<3)=\int_1^3 f(x)\,dx=\int_2^3 \frac{1}{2}dx=\frac{1}{2}.$$

Problem 4.23 If X_1, X_2, X_3, X_4 are (pairwise) uncorrelated random variables each having mean 0 and variance 1, compute the correlations of (1) X_1+X_2 and X_2+X_3; (2) X_1+X_2 and X_3+X_4.

Solution:

(1) $\text{Cov}(X_1+X_2, X_2+X_3)=\text{Cov}(X_1, X_2)+\text{Cov}(X_1, X_3)+\text{Cov}(X_2, X_2)+\text{Cov}(X_2, X_3)=\text{Cov}(X_2, X_2)=V(X_2)=1$.

$V(X_1+X_2)=2, V(X_2+X_3)=2$.

The correlation is $\frac{1}{\sqrt{2}\times\sqrt{2}}=\frac{1}{2}$.

(2) $\text{Cov}(X_1+X_2, X_3+X_4)=\text{Cov}(X_1, X_3)+\text{Cov}(X_1, X_4)+\text{Cov}(X_2, X_3)+\text{Cov}(X_2, X_4)=0$. The correlation is 0.

Problem 4.24 Let X and Y have a continuous distribution with joint PDF

$$f(x,y)=\begin{cases}x+y, & 0\leqslant x\leqslant 1 \text{ and } 0\leqslant y\leqslant 1\\ 0, & \text{otherwise}\end{cases}.$$

Compute the covariance $\text{Cov}(X,Y)$.

Solution:

The means of X and Y are the same since $f(x,y)=f(y,x)$ for all x and y. The mean of X (and the mean of Y) is

$$E(X)=\int_0^1\left[\int_0^1 x(x+y)\,\mathrm{d}x\right]\mathrm{d}y=\int_0^1\left(\frac{1}{3}+\frac{y}{2}\right)\mathrm{d}y=\frac{1}{3}+\frac{1}{4}=\frac{7}{12}.$$

Also,

$$E(XY)=\int_0^1\left[\int_0^1 xy(x+y)\,\mathrm{d}x\right]\mathrm{d}y=\int_0^1\left(\frac{y}{3}+\frac{y^2}{2}\right)\mathrm{d}y=\frac{1}{6}+\frac{1}{6}=\frac{1}{3}.$$

Thus, $\mathrm{Cov}(X,Y)=E(XY)-E(X)E(Y)=\frac{1}{3}-\left(\frac{7}{12}\right)^2=-0.006\,9.$

Problem 4.25 The random variables X and Y have a joint PDF given by

$$f(x,y)=\begin{cases}\dfrac{2\,\mathrm{e}^{-2x}}{x}, & 0<x<\infty \text{ and } 0\leqslant y<x\\ 0, & \text{otherwise}\end{cases}.$$

Compute the covariance $\mathrm{Cov}(X,Y)$.

Solution:

The support S of $f(x,y)$ is plotted in Figure 4.5.

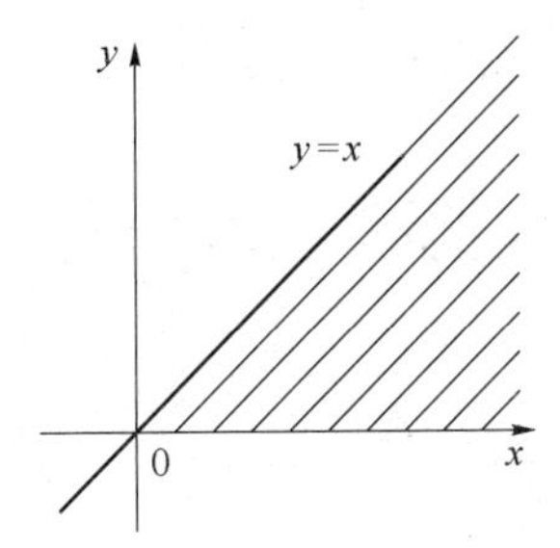

Figure 4.5

$$\begin{aligned}E(XY)&=\int_{-\infty}^{+\infty}\int_{-\infty}^{+\infty}xyf(x,y)\,\mathrm{d}x\mathrm{d}y\\&=\iint_S xy\times\frac{2\,\mathrm{e}^{-2x}}{x}\mathrm{d}x\mathrm{d}y\\&=\int_0^{+\infty}\left(\int_0^x 2y\mathrm{e}^{-2x}\mathrm{d}y\right)\mathrm{d}x\\&=\int_0^{+\infty}x^2\,\mathrm{e}^{-2x}\mathrm{d}x\\&=\frac{1}{4}.\end{aligned}$$

$$
\begin{aligned}
E(X) &= \int_{-\infty}^{+\infty}\int_{-\infty}^{+\infty} xf(x,y)\,\mathrm{d}x\mathrm{d}y \\
&= \iint_S x \times \frac{2\mathrm{e}^{-2x}}{x}\mathrm{d}x\mathrm{d}y \\
&= \int_0^{+\infty}\left(\int_0^x 2\,\mathrm{e}^{-2x}\mathrm{d}y\right)\mathrm{d}x \\
&= \int_0^{+\infty} 2x\,\mathrm{e}^{-2x}\mathrm{d}x \\
&= \frac{1}{2}.
\end{aligned}
$$

$$
\begin{aligned}
E(Y) &= \int_{-\infty}^{+\infty}\int_{-\infty}^{+\infty} yf(x,y)\,\mathrm{d}x\mathrm{d}y \\
&= \iint_S y \times \frac{2\,\mathrm{e}^{-2x}}{x}\mathrm{d}x\mathrm{d}y \\
&= \int_0^{+\infty}\left(\int_0^x \frac{2y\mathrm{e}^{-2x}}{x}\mathrm{d}y\right)\mathrm{d}x \\
&= \int_0^{+\infty}\frac{\mathrm{e}^{-2x}}{x}\left(\int_0^x 2y\mathrm{d}y\right)\mathrm{d}x \\
&= \int_0^{+\infty} x\mathrm{e}^{-2x}\mathrm{d}x \\
&= \frac{1}{4}.
\end{aligned}
$$

Therefore, $\mathrm{Cov}(X,Y)=E(XY)-E(X)E(Y)=\frac{1}{4}-\frac{1}{2}\times\frac{1}{4}=\frac{1}{8}$.

Problem 4.26 Suppose that X and Y have a continuous joint distribution for which the joint PDF is as follows:

$$
f(x,y)=\begin{cases}\frac{1}{3}(x+y), & 0\leqslant x\leqslant 1 \text{ and } 0\leqslant y\leqslant 2 \\ 0, & \text{otherwise}\end{cases}.
$$

Determine the value of $V(2X-3Y+8)$.

 **Solution**:

The support S of $f(x,y)$ is plotted in Figure 4.6.

$$E(X)=\int_0^1\left[\int_0^2 x\times\frac{1}{3}(x+y)\mathrm{d}y\right]\mathrm{d}x=\frac{5}{9}.$$

$$E(Y)=\int_0^1\left[\int_0^2 y\times\frac{1}{3}(x+y)\mathrm{d}y\right]\mathrm{d}x=\frac{11}{9}.$$

$$E(X^2)=\int_0^1\left[\int_0^2 x^2\times\frac{1}{3}(x+y)\mathrm{d}y\right]\mathrm{d}x=\frac{7}{18}.$$

$$E(Y^2)=\int_0^1\left[\int_0^2 y^2\times\frac{1}{3}(x+y)\mathrm{d}y\right]\mathrm{d}x=\frac{16}{9}.$$

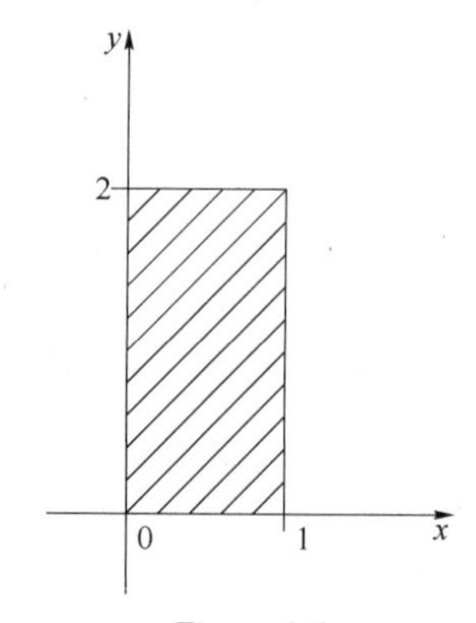

Figure 4.6

$$E(XY)=\int_0^1\left[\int_0^2 xy\cdot\frac{1}{3}(x+y)\mathrm{d}y\right]\mathrm{d}x=\frac{2}{3}.$$

Therefore,

$$V(X)=\frac{7}{18}-\left(\frac{5}{9}\right)^2=\frac{13}{162},$$

$$V(Y)=\frac{16}{9}-\left(\frac{11}{9}\right)^2=\frac{23}{81},$$

$$\mathrm{Cov}(X,Y)=\frac{2}{3}-\frac{5}{9}\times\frac{11}{9}=-\frac{1}{81}.$$

It follows that

$$V(2X-3Y+8)=4V(X)+9V(Y)-2\times2\times3\times\mathrm{Cov}(X,Y)=\frac{245}{81}.$$

Problem 4.27 Suppose that X and Y are random variables such that $V(X)=9, V(Y)=4$, and $\rho_{X,Y}=-\frac{1}{6}$. Determine $V(X+Y)$ and $V(X-3Y+4)$.

Solution:

$$\mathrm{Cov}(X,Y)=\rho_{X,Y}\sqrt{V(X)}\sqrt{V(Y)}=\left(-\frac{1}{6}\right)\times3\times2=-1.$$

$$V(X+Y)=V(X)+V(Y)+2\mathrm{Cov}(X,Y)=11.$$

$$V(X-3Y+4)=V(X)+9V(Y)-2\times3\times\mathrm{Cov}(X,Y)=51.$$

Problem 4.28 Suppose that an automobile dealer pays an amount X (in thousands of dollars) for a used car and then sells it for an amount Y. Suppose that the random variables X and Y have the following joint PDF:

$$f(x,y)=\begin{cases}\frac{1}{36}x, & 0<x<y<6\\ 0, & \text{otherwise}\end{cases}.$$

Determine the dealer's expected gain from the sale.

Solution:

The support S of $f(x,y)$ is plotted in Figure 4.7.

The dealer's expected gain is

$$E(Y-X)=\iint_S(y-x)f(x,y)\mathrm{d}x\mathrm{d}y=\int_0^6\left[\int_0^y(y-x)\frac{1}{36}x\mathrm{d}x\right]\mathrm{d}y=\frac{3}{2}.$$

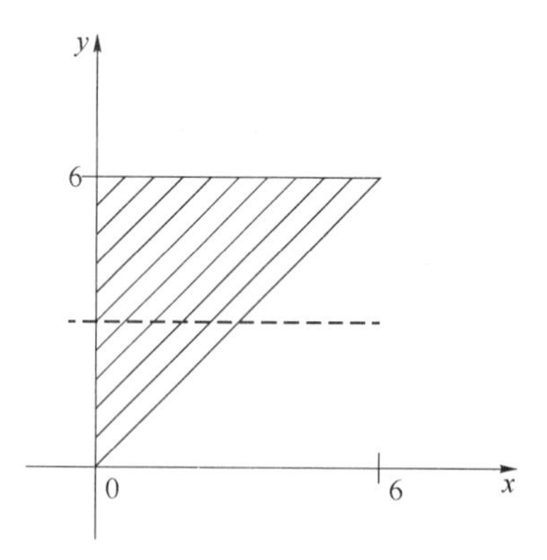

Figure 4.7

Problem 4.29 Four independent random variables X_1, X_2, X_3 and X_4 each normal distributed with a mean of 0 and a standard deviation of 4 are combined to form a new random variable

$$Y=X_1^2+X_2^2+X_3^2+X_4^2.$$

(1) What is the expected value of Y?

(2) What is the variance of Y?

Solution:

(1) For $i=1,2,3,4, E(X_i)=0, V(X_i)=4^2=16$. Then $E(X_i^2)=V(X_i)+[E(X_i)]^2=16$.

$$E(Y)=E(X_1^2+X_2^2+X_3^2+X_4^2)=\sum_{i=1}^{4}E(X_i^2)=64.$$

(2) Suppose random variable Z follows the normal distribution $N(0,1)$, then $X_i=4Z$. And it is known that $E(Z^4)=3$. So $E(X_i^4)=E[(4Z)^4]=4^4E(Z^4)=768. V(X_i^2)=E(X_i^4)-[E(X_i^2)]^2=768-16^2=512$. Because the random variables X_1, X_2, X_3 and X_4 are independent,

$$V(Y)=V(X_1^2+X_2^2+X_3^2+X_4^2)=\sum_{i=1}^{4}V(X_i^2)=2\,048.$$

Note: For $Z \sim N(0,1)$, if n is odd, $E(Z^n=0)$; If n is even, $E(Z^n)=(n-1)\times(n-3)\times\cdots\times 1$.

Problem 4.30 Let X and Y be random variables with $E(X)=2, V(X)=1, E(Y)=3, V(Y)=4$. What are the smallest and largest possible values of $V(X+Y)$?

Solution:

$$V(X+Y)=V(X)+V(Y)+2\text{Cov}(X,Y)$$

$$|\rho_{X,Y}|\leqslant 1, -\sqrt{V(X)}\sqrt{V(Y)}\leqslant \text{Cov}(X,Y)\leqslant \sqrt{V(X)}\sqrt{V(Y)}$$

Therefore,

$$V(X)+V(Y)-2\sqrt{V(X)}\sqrt{V(Y)}\leqslant V(X+Y)\leqslant V(X)+V(Y)+2\sqrt{V(X)}\sqrt{V(Y)}.$$

So the smallest and largest possible values of $V(X+Y)$ are 1 and 9.

Problem 4.31 Your company must make a sealed bid for a construction project. If you succeed in winning the contract (by having the lowest bid), then you plan to pay another firm \$100,000 to do the work. If you believe that the minimum bid (in thousands of dollars) of the other participating companies can be modeled as being the value of a random variable that is uniformly distributed on (70, 140), how much should you bid to maximize your expected profit?

 **Solution:**

Let X be the minimum bid (in thousands of dollars) of the other participating companies. $X \sim U(70,140)$. The PDF of X is as follows:

$$f_X(x)=\begin{cases}\frac{1}{70}, & 70<x<140 \\ 0, & \text{otherwise}\end{cases}.$$

Let a be your bid and Y be your profit. Then

(1) if $a \leqslant 70$, you win the bid and the profit is $a-100<0$.

(2) if $a \geqslant 140$, you lose the bid and the profit is 0.

(3) if $70<a<140$,

$$Y=r(X)=\begin{cases}a-100, & a<X \\ 0, & a \geqslant X\end{cases}.$$

$$\begin{aligned}E(Y) &= \int_{-\infty}^{+\infty} r(x) f_X(x)\,\mathrm{d}x \\ &= \int_{70}^{140} r(x) \times \frac{1}{70}\mathrm{d}x \\ &= \int_{70}^{a} 0 \times \frac{1}{70}\mathrm{d}x + \int_{a}^{140} (a-100) \times \frac{1}{70}\mathrm{d}x \\ &= \frac{1}{70}(-a^2+240a-14\,000).\end{aligned}$$

$E(Y)$ is a function of a. Let $\frac{\mathrm{d}}{\mathrm{d}a}E(Y)=\frac{1}{35}(-a+120)=0, a=120$. $\frac{\mathrm{d}^2}{\mathrm{d}a^2}E(Y)=-\frac{1}{35}<0$. Therefore, the profit achieves its maximum at $a=120$ (in thousands of dollars). The profit is $E(Y)=\frac{40}{7}$ (in thousands of dollars).

Problem 4.32 Two independent random variables X and Y have the variances $V(X)=12$ and $V(Y)=18$. Let $W=-4X+2Y$ and $Z=3X-6Y$.

(1) Find the variances of W and Z.

(2) Find the correlation coefficient of W and Z.

 **Solution:**

(1) Since X and Y are independent,

$$V(W)=V(-4X+2Y)=16V(X)+4V(Y)=264$$

$$V(Z)=V(3X-6Y)=9V(X)+36V(Y)=756.$$

(2) $\text{Cov}(W,Z)=\text{Cov}(-4X+2Y,3X-6Y)=-12[V(X)+V(Y)]=-360$. The correlation coefficient of W and Z is

$$\rho_{W,Z}=\frac{\text{Cov}(W,Z)}{\sqrt{V(W)}\sqrt{V(Z)}}=\frac{-360}{\sqrt{264}\sqrt{756}}=-0.805\,8.$$

Problem 4.33 Find the mean and variance of the distribution whose PDF is given as

$$f(x)=\frac{1}{\sqrt{128\pi}}e^{-\frac{(x+7)^2}{128}}.$$

 **Solution:**

The PDF can be rewritten as

$$f(x)=\frac{1}{\sqrt{2\pi}\times 8}e^{-\frac{[x-(-7)]^2}{2\times 8^2}}.$$

The random variable follows the normal distribution with parameters $\mu=-7,\sigma=8$. Thus, the mean is -7 and the variance is $8^2=64$.

Problem 4.34 Find the mean and variance of the distribution whose PDF is given as

$$f(x)=\frac{1}{\sqrt{\pi}}e^{-x^2+2x-1}.$$

Solution:

The PDF can be rewritten as

$$f(x)=\frac{1}{\sqrt{2\pi}\times\frac{\sqrt{2}}{2}}e^{-\frac{(x-1)^2}{2\times\left(\frac{\sqrt{2}}{2}\right)^2}}.$$

The random variable follows the normal distribution with parameters $\mu=1,\sigma=\frac{\sqrt{2}}{2}$. Thus, the mean is 1 and the variance is $\left(\frac{\sqrt{2}}{2}\right)^2=\frac{1}{2}$.

Problem 4.35 Let X be a 2×1 random vector and denote its components by X_1 and X_2. The covariance matrix of X is

$$V[\boldsymbol{X}]=\begin{bmatrix}4 & 1\\1 & 2\end{bmatrix}.$$

Compute the variance of the random variable $Y=3X_1+4X_2$.

Solution:

$V(Y)=V(3X_1+4X_2)=9V(X_1)+16V(X_2)+2\times3\times4\times\mathrm{Cov}(X_1,X_2)=9\times4+16\times2+24\times1=92.$

Problem 4.36 In n independent Bernoulli trials, each with probability of success p, let X be the number of successes and let Y be the number of failures. Calculate $E(XY)$ and $\mathrm{Cov}(X,Y)$.

Solution:

X denotes the number of successes and Y denotes the number of failures, then $Y=n-X$, $X\sim \mathrm{B}(n,p)$, $E(X)=np$, $V(X)=np(1-p)$, $E(X^2)=np(1-p)+n^2p^2$.

$$E(XY)=E[X(n-X)]=nE(X)-E(X^2)=n(n-1)p(1-p).$$

$$E(Y)=E(n-X)=n(1-p).$$

$\mathrm{Cov}(X,Y)=E(XY)-E(X)E(Y)=n(n-1)p(1-p)-np\times n(1-p)=-np(1-p).$

Or

$$\mathrm{Cov}(X,Y)=\mathrm{Cov}(X,n-X)=-V(X)=-np(1-p).$$

Problem 4.37 Use the Chebyshev's inequality to find an upper bound on the probability that the number of tails that come up when a fair coin is tossed n times deviates from the mean by more than $5\sqrt{n}$.

Solution:

Let X denote the number of tails, then $X\sim \mathrm{B}\left(n,\dfrac{1}{2}\right)$. We want to determine an upper bound on $P(|X-E(X)|\geqslant 5\sqrt{n})$.

According to the Chebyshev's inequality,

$$P[|X-E(X)|\geqslant 5\sqrt{n}]\leqslant\frac{V(X)}{(5\sqrt{n})^2}=\frac{n\times\frac{1}{2}\times\left(1-\frac{1}{2}\right)}{25n}=\frac{1}{100}.$$

Problem 4.38 The joint PDF of (X,Y) is given by

$$f(x,y)=\begin{cases}\dfrac{1}{4}(1-x^3y+xy^3), & -1<x<1,-1<y<1\\ 0, & \text{otherwise}\end{cases}.$$

Show that X and Y are uncorrelated and dependent.

Solution:

The support $S=\{(x,y)\mid -1<x<1,-1<y<1\}$ of $f(x,y)$ is plotted in Figure 4.8.

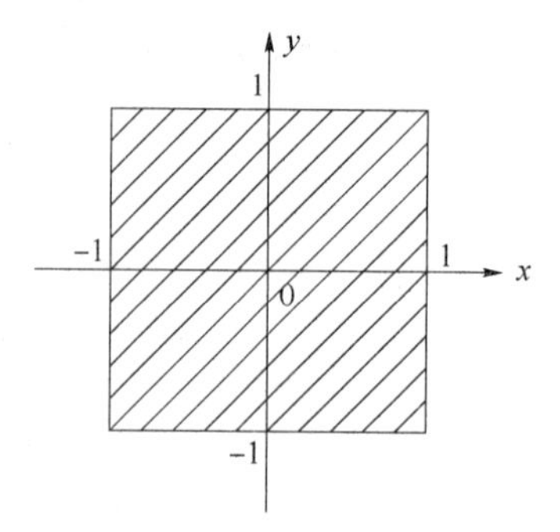

Figure 4.8

Find the marginal PDF of X.

- If $x\leqslant -1$ or $x\geqslant 1$, $f_X(x)=0$.
- If $-1<x<1$, $f_X(x)=\int_{-\infty}^{+\infty} f(x,y)\,\mathrm{d}y=\int_{-1}^{1}\frac{1}{4}(1-x^3y-xy^3)\,\mathrm{d}y=\frac{1}{2}$.

The marginal PDF of X is

$$f_X(x)=\begin{cases}\frac{1}{2}, & -1<x<1\\ 0, & \text{otherwise}\end{cases}.$$

$$E(X)=\int_{-\infty}^{+\infty} x f_X(x)\,\mathrm{d}x=\int_{-1}^{1}\frac{x}{2}\mathrm{d}x=0.$$

Find the marginal PDF of Y.

- If $y\leqslant -1$ or $y\geqslant 1$, $f_Y(y)=0$.
- If $-1<y<1$, $f_Y(y)=\int_{-\infty}^{+\infty} f(x,y)\,\mathrm{d}x=\int_{-1}^{1}\frac{1}{4}(1-x^3y-xy^3)\,\mathrm{d}x=\frac{1}{2}$.

The marginal PDF of Y is

$$f_Y(y)=\begin{cases}\frac{1}{2}, & -1<y<1\\ 0, & \text{otherwise}\end{cases}.$$

$$E(Y)=\int_{-\infty}^{+\infty} y f_Y(y)\,\mathrm{d}x=\int_{-1}^{1}\frac{y}{2}\mathrm{d}y=0.$$

$$\begin{aligned}E(XY)&=\int_{-\infty}^{+\infty}\int_{-\infty}^{+\infty} xyf(x,y)\,\mathrm{d}x\mathrm{d}y\\&=\int_{-1}^{1}\left[\int_{-1}^{1}\frac{1}{4}xy(1-x^3y-xy^3)\,\mathrm{d}y\right]\mathrm{d}x\\&=\int_{-1}^{1}\left(\frac{x^2}{10}-\frac{x^4}{6}\right)\mathrm{d}x\end{aligned}$$

$$= 2\int_0^1 \left(\frac{x^2}{10} - \frac{x^4}{6}\right) dx$$

$$= 0.$$

$$\operatorname{Cov}(X,Y) = E(XY) - E(X)E(Y) = 0.$$

Therefore, X and Y are uncorrelated.

In the support set, $f(x,y) \neq f_X(x)f_Y(y)$, so that X and Y are dependent.

Problem 4.39 The joint PF of (X,Y) is given by

	$X=-2$	$X=-1$	$X=1$	$X=2$
$Y=1$	0	$\frac{1}{4}$	$\frac{1}{4}$	0
$Y=4$	$\frac{1}{4}$	0	0	$\frac{1}{4}$

Show that X and Y are uncorrelated and dependent.

 **Solution:**

$$E(XY) = (-1)\times1\times\frac{1}{4} + 1\times1\times\frac{1}{4} + (-2)\times4\times\frac{1}{4} + 2\times4\times\frac{1}{4} = 0.$$

The marginal PF of X is

	$X=-2$	$X=-1$	$X=1$	$X=2$
P	$\frac{1}{4}$	$\frac{1}{4}$	$\frac{1}{4}$	$\frac{1}{4}$

$$E(X) = (-2)\times\frac{1}{4} + (-1)\times\frac{1}{4} + 1\times\frac{1}{4} + 2\times\frac{1}{4} = 0.$$

The marginal PF of Y is

	$Y=1$	$Y=4$
P	$\frac{1}{2}$	$\frac{1}{2}$

$$E(Y) = 1\times\frac{1}{2} + 4\times\frac{1}{2} = \frac{5}{2}.$$

$$\operatorname{Cov}(X,Y) = E(XY) - E(X)E(Y) = 0.$$

So X and Y are uncorrelated. But

$0 = P(X=-2, Y=1) \neq P(X=-2)P(Y=1) = \frac{1}{4}\times\frac{1}{2}$, X and Y are dependent.

Problem 4.40 Suppose A and B are two events. $P(A)>0$, $P(B)>0$, Let

$$X=\begin{cases}1, & A \text{ occurs}\\ -1, & A \text{ does not occur}\end{cases}, \quad Y=\begin{cases}1, & B \text{ occurs}\\ -1, & B \text{ does not occur}\end{cases}.$$

Show that if $\rho_{X,Y}=0$, then X and Y are independent.

Solution:

$$
\begin{aligned}
E(XY) &= 1\times1\times P(X=1,Y=1)+1\times(-1)\times P(X=1,Y=-1) \\
&\quad +(-1)\times1\times P(X=-1,Y=1)+(-1)\times(-1)\times P(=-1,Y=-1) \\
&= P(AB)-P(AB^c)-P(A^cB)+P(A^cB^c) \\
&= P(AB)-[P(A)-P(AB)]-[P(B)-P(AB)]+1-P(A)-P(B)+P(AB) \\
&= 4P(AB)-2P(A)-2P(B)+1.
\end{aligned}
$$

$E(X)=1\times P(A)+(-1)\times P(A^c)=2P(A)-1$

$E(Y)=1\times P(B)+(-1)\times P(B^c)=2P(B)-1$

$E(X)E(Y)=[2P(A)-1]\times[2P(B)-1]=4P(A)P(B)-2P(A)-2P(B)+1.$

If $\rho_{X,Y}=0$, $\mathrm{Cov}(X,Y)=E(XY)-E(X)E(Y)=0$, $P(AB)=P(A)P(B)$.

The events A and B are independent, then A and B^c, A^c and B, A^c and B^c are independent.

$P(X=1,Y=1)=P(AB)=P(A)P(B)=P(X=1)P(Y=1).$

$P(X=1,Y=-1)=P(AB^c)=P(A)P(B^c)=P(X=1)P(Y=-1).$

$P(X=-1,Y=1)=P(A^cB)=P(A^c)P(B)=P(X=-1)P(Y=1).$

$P(X=-1,Y=-1)=P(A^cB^c)=P(A^c)P(B^c)=P(X=-1)P(Y=-1).$

Hence, X and Y are independent.

Chapter 5 Large Random Samples

Summary of Knowledge

Exercise Solutions

Problem 5.1 A device has ten parts. The lengths of the parts are IID. The mean of each part is 2 mm and the standard deviation is 0.05 mm. The device is non-defective if the total length is 20±0.1 mm. Find the probability that the device is non-defective.

Solution:

Let $X_k(k=1,\cdots,10)$ be the length of the k-th part. $E(X_k)=2, \sigma_{X_k}=0.05(k=1,\cdots,10)$.

The total length is $X = \sum_{k=1}^{10} X_k, E(X) = 20, \sigma_X = \sqrt{10} \times 0.05 = 0.158$.

By the central limit theorem, $\frac{X-20}{0.158} \dot{\sim} N(0,1)$.

$$
\begin{aligned}
P(20-0.1<X<20+0.1) &= P\left(-\frac{0.1}{0.158}<\frac{X-20}{0.158}<\frac{0.1}{0.158}\right) \\
&\approx P(-0.63<Z<0.63) \\
&= 2\Phi(0.63)-1 \\
&= 0.4714,
\end{aligned}
$$

where $Z \sim N(0,1)$.

Problem 5.2 Suppose the lifetime in hours of a certain type of electronic units follows exponnetial distribution with $\lambda=0.1$. Let T be the total lifetime of 30 electronic units of this type. Determine $P(T \geqslant 350)$.

Solution:

Let $X_k(k=1,\cdots,30)$ be the lifetime of the k-th unit. $E(X_k)=\frac{1}{\lambda}=10, \sigma_{X_k}=\sqrt{\frac{1}{\lambda^2}}=10$

$(k=1,\cdots,30)$. The total lifetime is $T=\sum_{k=1}^{30} X_k$, $E(T)=300$, $\sigma_T=\sqrt{30}\times 10=54.772$.

By the central limit theorem, $\frac{T-300}{54.772}\dot{\sim} N(0,1)$.

$$
\begin{aligned}
P(T\geqslant 350) &= P\left(\frac{T-300}{54.772}\geqslant\frac{350-300}{54.772}\right)\\
&\approx P(Z\geqslant 0.91)\\
&=1-\Phi(0.91)\\
&=0.181\ 4,
\end{aligned}
$$

where $Z\sim N(0,1)$.

Problem 5.3 Suppose that 75 percent of the people in a certain metropolitan area live in the city and 25 percent of the people live in the suburbs. If 1 200 people attending a certain concert represent a random sample from the metropolitan area, what is the probability that the number of people from the suburbs attending the concert will be fewer than 270?

Solution:

Let X denote the number of people in the sample that are from the suburbs. Then $X\sim B(1\ 200, 0.25)$. $E(X)=1\ 200\times 0.25=300$, $V(X)=1\ 200\times 0.25\times 0.75=225$.

By the De Moivre-Laplace Theorem, X can be approximated by $N(300,225)$.

$$
\begin{aligned}
P(X<270) &= P(X\leqslant 269.5)\quad \text{(continuity correction)}\\
&= P\left(\frac{X-300}{15}\leqslant\frac{269.5-300}{15}\right)\\
&\approx P(Z\leqslant -2.03)\\
&=1-\Phi(2.03)\\
&=0.021\ 2,
\end{aligned}
$$

where $Z\sim N(0,1)$.

Problem 5.4 Suppose that the distribution of the number of defects on any given bolt of cloth is the Poisson distribution with mean 5, and the number of defects on each bolt is counted for a random sample of 125 bolts. Determine the probability that the average number of defects per bolt in the sample will be less than 5.5.

Solution:

Suppose that X_i is the number of defects on the i-th bolt. Each X_i is a Poisson random variable with $\lambda=5$ (since the mean is 5 which is equal to λ). The standard deviation for each X_i is $\sqrt{\lambda}=\sqrt{5}$. If we assume that the number of defects on each bolt is independent of the number on any other bolt then the X_i's are a simple random sample with $n=125$.

By the Central Limit Theorem, the average number of bolts is approximately normal with $\mu_{\overline{X}}=5$ and the standard deviation is $\sigma_{\overline{X}}=\frac{\sqrt{5}}{\sqrt{125}}=\frac{1}{5}$. We then have

$$P(\bar{X}<5.5)=P\left(\frac{\bar{X}-5}{\frac{1}{5}}<\frac{5.5-5}{\frac{1}{5}}\right)\approx P(Z<2.5)=\Phi(2.5)=0.993\ 8,$$

where $Z\sim N(0,1)$.

Problem 5.5 Let X be the length of a pregnancy in days. Suppose that it has approximately a normal distribution with a mean of 266 days and a standard deviation of 16 days.

(1) What is the probability that a randomly selected pregnancy lasts more than 274 days?

(2) Suppose that we have a random sample of $n=25$ pregnant women. What is the probability that the sample mean pregnancy length is more than 274 days?

Solution:

(1) $X \dot{\sim} N(266,16^2)$.

$$\begin{aligned}P(X>274)&=P\left(\frac{X-266}{16}>\frac{274-266}{16}\right)\\&\approx P(Z>0.5)\\&=1-\Phi(0.5)\\&=1-0.691\ 5\\&=0.308\ 5,\end{aligned}$$

where $Z\sim N(0,1)$.

(2) By the central limit theorem, We know that $\bar{X}$ is normally distributed with mean 266 and standard deviation $16/\sqrt{25}=3.2$.

$$\begin{aligned}P(\bar{X}>274)&=P\left(\frac{\bar{X}-266}{3.2}>\frac{274-266}{3.2}\right)\\&\approx P(Z>2.5)\\&=1-\Phi(2.5)\\&=1-0.993\ 8\\&=0.006\ 2,\end{aligned}$$

where $Z\sim N(0,1)$.

Problem 5.6 Suppose that a random sample of size n is to be taken from a distribution for which the mean is μ and the standard deviation is 3. Use the central limit theorem to determine approximately the smallest value of n for which the following relation will be satisfied:

$$P(|\bar{X}-\mu|<0.3)\geqslant 0.95.$$

Solution:

By the central limit theorem, $\bar{X} \dot{\sim} N\left(\mu,\frac{3^2}{n}\right)$. The distribution of $\frac{\bar{X}-\mu}{3/\sqrt{n}}$ will be approximately the standard normal distribution $Z\sim N(0,1)$. Therefore,

$$P(|\overline{X}-\mu|<0.3)\approx P(|Z|<0.1\sqrt{n})=2\Phi(0.1\sqrt{n})-1.$$

But $2\Phi(0.1\sqrt{n})-1\geqslant 0.95$ if and only if $\Phi(0.1\sqrt{n})\geqslant\frac{1+0.95}{2}=0.975$, and this inequality is satisfied if and only if $0.1\sqrt{n}\geqslant 1.96$, or equivalently, $n\geqslant 384.16$. Hence, the smallest possible value of n is 385.

Problem 5.7 Suppose that the proportion of defective items in a large manufactured lot is 0.1. What is the smallest random sample of items that must be taken from the lot in order for the probability to be at least 0.99 that the proportion of defective items in the sample will be less than 0.13?

Solution:

By the central limit theorem, the distribution of the proportion $\overline{X}$ of defective items in the sample will be approximately the normal distribution with mean $E(\overline{X})=0.1$ and variance $V(\overline{X})=\frac{1}{n^2}(n\times 0.1\times 0.9)=\frac{0.09}{n}$.

Therefore, the distribution of $\sqrt{n}(\overline{X}-0.1)/0.3$ will be approximately the standard normal distribution $Z\sim N(0,1)$. It follows that

$$P(\overline{X}<0.13)\approx P(Z<0.1\sqrt{n})=\Phi(0.1\sqrt{n}).$$

For this value to be at least 0.99, we must have $0.1\sqrt{n}\geqslant 2.33$ or, equivalently, $n\geqslant 542.89$. Hence, the smallest possible value of n is 543.

Problem 5.8 A local bakery randomly sells three types of cupcakes. Accordingly, the cupcake's price in dollars is a random variable with the following PF.

X	1	1.2	1.5
P	0.3	0.2	0.5

On a given day, the bakery sold 300 cupcakes.

(1) Find the probability that its earning is at least 400 dollars.

(2) Find the probability that the bakery sold at least 60 cupcakes at the price of 1.2 dollars.

Solution:

(1) Let X_i be the price of the i-th cupcake, $i=1,2,\cdots,300$.

$$E(X_i)=1\times 0.3+1.2\times 0.2+1.5\times 0.5=1.29$$

$$E(X_i^2)=1^2\times 0.3+1.2^2\times 0.2+1.5^2\times 0.5=1.713$$

$$V(X_i)=E(X_i^2)-[E(X_i)]^2=0.048\ 9.$$

Let X denote the earning, $X=\sum_{i=1}^{300} X_i, E(X)=300\times 1.29=387, \sigma_X=\sqrt{300}\times\sqrt{0.048\ 9}\approx$ 3.83.

By the central limit theorem, the distribution of $\frac{X-387}{3.83}$ will be approximately the standard normal distribution $Z\sim \mathrm{N}(0,1)$. Therefore,

$$\begin{aligned}P(X\geqslant 400)&=1-P(X<400)\\&=1-P\left(\frac{X-387}{3.83}<\frac{400-387}{3.83}\right)\\&\approx 1-P(Z<3.39)\\&=1-\Phi(3.39)\\&=0.000\ 3.\end{aligned}$$

(2) Let Y denote the number of cupcakes at the price of 1.2 dollars, then $Y\sim \mathrm{B}(300,0.2)$. By the De Moivre-Laplace theorem, Y can be approximated by $\mathrm{N}(60,48)$.

$$\begin{aligned}P(Y\geqslant 60)&=P(Y\geqslant 59.5)\quad \text{(continuity correction)}\\&=P\left(\frac{Y-300\times 0.2}{\sqrt{300\times 0.2\times 0.8}}\geqslant\frac{59.5-300\times 0.2}{\sqrt{300\times 0.2\times 0.8}}\right)\\&\approx P(Z\geqslant -0.07)\\&=\Phi(0.07)\\&=0.527\ 9.\end{aligned}$$

Problem 5.9 A physicist makes 25 independent measurements of the specific gravity of a certain body. He knows that the limitations of his equipment are such that the standard deviation of each measurement is σ units.

(1) By using the Chebyshev's inequality, find a lower bound for the probability that the average of his measurements will differ from the actual specific gravity of the body by less than $\sigma/4$ units.

(2) By using the central limit theorem, find an approximate value for the probability in part(1).

Solution:

(1) Let $\overline{X}$ denote the average of his measurements. $V(\overline{X})=\frac{\sigma^2}{25}$. By the Chebyshev's inequality,

$$P\left(|\overline{X}-\mu|\geqslant\frac{\sigma}{4}\right)\leqslant\frac{\frac{\sigma^2}{25}}{\left(\frac{\sigma}{4}\right)^2}=\frac{16}{25}.$$

Therefore,

$$P\left(|\bar{X}-\mu|<\frac{\sigma}{4}\right)\geqslant 1-\frac{16}{25}=0.36.$$

(2) The distribution of $\frac{\bar{X}-\mu}{\sigma/\sqrt{n}}=\frac{5}{\sigma}(\bar{X}-\mu)$ will be approximately the standard normal distribution $Z\sim N(0,1)$. Therefore,

$$P\left(|\bar{X}-\mu|\leqslant\frac{\sigma}{4}\right)\approx P\left(|Z|\leqslant\frac{5}{4}\right)=2\Phi(1.25)-1=0.788\ 7.$$

Problem 5.10 Let $X_1,\cdots,X_{30}$ be independent random variables each having a discrete distribution with PF

$$f(x)=\begin{cases}\frac{1}{4}, & x=0 \text{ or } 2\\ \frac{1}{2}, & x=1\\ 0, & \text{otherwise}\end{cases}.$$

Use the central limit theorem and the correction for continuity to approximate the probability that $X_1+\cdots+X_{30}$ is at most 33.

Solution:

For $i=1,2,\cdots,30$,

$$E(X_i)=0\times\frac{1}{4}+1\times\frac{1}{2}+2\times\frac{1}{4}=1.$$

$$E(X_i^2)=0^2\times\frac{1}{4}+1^2\times\frac{1}{2}+2^2\times\frac{1}{4}=\frac{3}{2}.$$

$$V(X_i)=E(X_i^2)-[E(X_i)]^2=\frac{1}{2}.$$

By the central limit theorem, $Y=X_1+\cdots+X_{30}$ follows approximately the normal distribution with mean 30 and variance 15. Using the correction for continuity,

$$\begin{aligned}P(Y\leqslant 33)&=P(Y<33.5)\\&=P\left(\frac{Y-30}{\sqrt{15}}<\frac{33.5-30}{\sqrt{15}}\right)\\&\approx P(Z<0.9)\\&=\Phi(0.9)\\&=0.815\ 9,\end{aligned}$$

where $Z\sim N(0,1)$.

Problem 5.11 Suppose that a pair of balanced dice are rolled 120 times, and let X denote the number of rolls on which the sum of the two numbers is 7. Use the central limit theorem to determine a value of k such that $P(|X-20|\leqslant k)$ is approximately 0.95.

Solution:

$X \sim B\left(120, \frac{1}{6}\right)$. $E(X) = np = 120 \times \frac{1}{6} = 20$. $V(X) = np(1-p) = 120 \times \frac{1}{6} \times \frac{5}{6} = 16.666\ 7$. $\sigma_X = \sqrt{V(X)} = 4.082$.

By using the central limit theorem, $X \stackrel{\cdot}{\sim} N(20, 16.666\ 7)$. $\frac{X-20}{4.082} \stackrel{\cdot}{\sim} N(0,1)$.

$P(|X-20| \leqslant k) = P\left(\frac{|X-20|}{4.082} \leqslant \frac{k}{4.082}\right) \approx P\left(|Z| \leqslant \frac{k}{4.082}\right) = 2\Phi\left(\frac{k}{4.082}\right) - 1$, where $Z \sim N(0,1)$.

Therefore, $2\Phi\left(\frac{k}{4.082}\right) - 1 = 0.95$, $\Phi\left(\frac{k}{4.082}\right) = 0.975$. Using the standard normal table, $\frac{k}{4.082} = 1.96$, $k = 8$.

Problem 5.12 Let X be normally distributed with standard deviation σ. Determine $P(|X-\mu| \geqslant 2\sigma)$. Compare with Chebyshev's inequality.

Solution:

The probability is

$$P(|X-\mu| \geqslant 2\sigma) = P(|Z| \geqslant 2) = 2 - 2\Phi(2) = 0.045\ 6, \text{ where } Z \sim N(0,1).$$

In contrast, Chebyshev's inequality only gives the weaker statement

$$P(|X-\mu| \geqslant 2\sigma) \leqslant \frac{\sigma^2}{(2\sigma)^2} = 0.25.$$

Problem 5.13 Let X be a random variable with $E(X) = 0$ and $V(X) = 1$. What integer value k will assure us that $P(|X| \geqslant k) \leqslant 0.01$?

Solution:

By the Chebyshev's inequality,

$$P(|X-E(X)| \geqslant k) \leqslant \frac{V(X)}{k^2}.$$

So

$$P(|X| \geqslant k) \leqslant \frac{1}{k^2}.$$

We take $k = 10$ to meet the requirement.

Problem 5.14 A certain drug is to be rated either effective or ineffective. Suppose lab results indicate that 75% of the time the drug increases the lifespan of a patient by 5 years (effective) and 25% of the time the drug causes a complication which decreases the lifespan of a patient by 1 year(ineffective). As part of a study, you administer the drug to 10 000 patients. What is the probability that the lifespans of those in the study will be increased by an average of 3.55 years or more?

 **Solution:**

Let $X_i(i=1,\cdots,10\ 000)$ denote the increased lifespan of the i-th patient.

$$E(X_i)=\mu=5\times0.75-1\times0.25=3.5.$$

$$E(X_i^2)=5^2\times0.75+(-1)^2\times0.25=19.$$

$$V(X_i)=E(X_i^2)-[E(X_i)]^2=6.75.$$

$$\sigma=\sqrt{V(X_i)}=\sqrt{6.75}\approx2.598\ 1.$$

By the central limit theorem, the averaged increased lifespan follows the normal distribution with $E(\overline{X})=\mu=3.5$, and $\sigma_{\overline{X}}=\dfrac{\sigma}{\sqrt{n}}=0.01\times\sqrt{6.75}\approx 0.026$.

$$\begin{aligned}P(\overline{X}\geqslant3.55)&=P\left(\frac{\overline{X}-\mu}{\frac{\sigma}{\sqrt{n}}}\geqslant\frac{3.55-3.5}{\frac{\sqrt{6.75}}{100}}\right)\\&\approx P(Z\geqslant1.92)\\&=1-P(Z<1.92)\\&=1-0.972\ 6\\&=0.027\ 4,\end{aligned}$$

where $Z\sim N(0,1)$.

Chapter 6 Estimation

Summary of Knowledge

Exercise Solutions

Problem 6.1 Suppose $X \sim N(\mu, \sigma^2)$, where μ is known and σ^2 is unknown. $(X_1, X_2, \cdots, X_n)$ is a sample from this distribution.

(1) Determine the joint PDF of $(X_1, X_2, \cdots, X_n)$;

(2) Identify the statistics from the following functions. $\sum_{i=1}^{n} X_i^2$, $\max_{1 \leqslant i \leqslant n} \{X_i\}$, $\frac{\sum_{i=1}^{n} (X_i - \mu)^2}{\sigma^2}$, $\sum_{i=1}^{n} X_i + \mu$.

Solution:

(1) The joint PDF of $(X_1, X_2, \cdots, X_n)$ is

$$f(x_1, x_2, \cdots, x_n) = \prod_{i=1}^{n} \frac{1}{\sqrt{2\pi}\sigma} e^{-\frac{(x_i-\mu)^2}{2\sigma^2}} = (2\pi)^{-\frac{n}{2}} \sigma^{-n} e^{-\frac{1}{2\sigma^2}\sum_{i=1}^{n}(x_i-\mu)^2}.$$

(2) $\sum_{i=1}^{n} X_i^2$, $\max_{1 \leqslant i \leqslant n} \{X_i\}$, $\sum_{i=1}^{n} X_i + \mu$ are statistics, but $\frac{\sum_{i=1}^{n} (X_i - \mu)^2}{\sigma^2}$ is not a statistic, because it contains the unknown parameter σ^2.

Problem 6.2 Suppose we have the following sample of $n = 15$ observations: 0.22, −0.67, −0.07, 1.65, 0.66, 0.88, 0.74, 0.72, −1.19, 0.14, −1.02, −0.21, −0.54, 0.99, 1.39. Find the empirical CDF and sketch its plot.

Solution:

Sort the observations from the smallest to the largest, −1.19, −1.02, −0.67, −0.54, −0.21,

−0.07, 0.14, 0.22, 0.66, 0.72, 0.74, 0.88, 0.99, 1.39, 1.65. The empirical CDF is

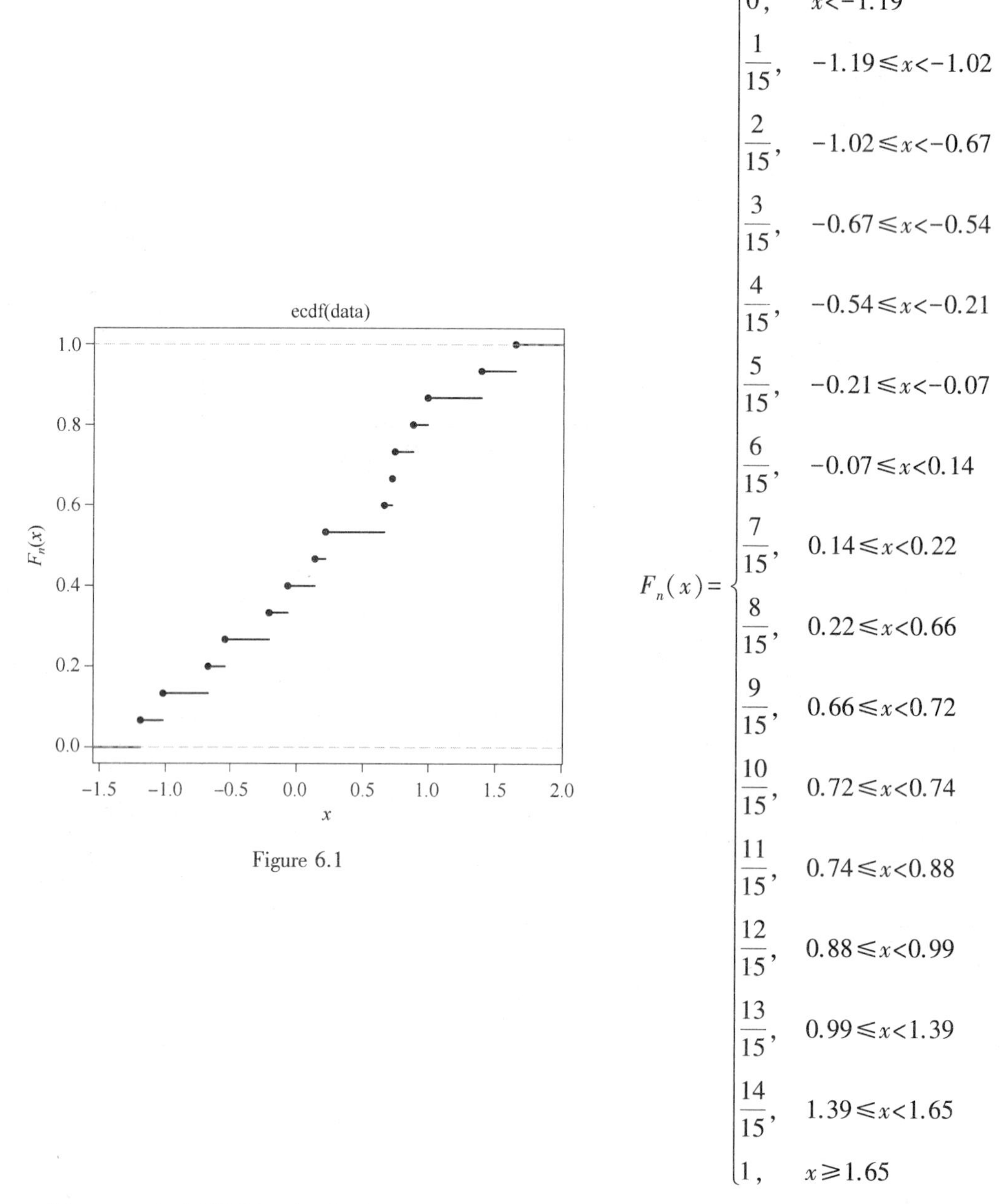

Figure 6.1

$$F_n(x)=\begin{cases}0, & x<-1.19\\ \frac{1}{15}, & -1.19\leqslant x<-1.02\\ \frac{2}{15}, & -1.02\leqslant x<-0.67\\ \frac{3}{15}, & -0.67\leqslant x<-0.54\\ \frac{4}{15}, & -0.54\leqslant x<-0.21\\ \frac{5}{15}, & -0.21\leqslant x<-0.07\\ \frac{6}{15}, & -0.07\leqslant x<0.14\\ \frac{7}{15}, & 0.14\leqslant x<0.22\\ \frac{8}{15}, & 0.22\leqslant x<0.66\\ \frac{9}{15}, & 0.66\leqslant x<0.72\\ \frac{10}{15}, & 0.72\leqslant x<0.74\\ \frac{11}{15}, & 0.74\leqslant x<0.88\\ \frac{12}{15}, & 0.88\leqslant x<0.99\\ \frac{13}{15}, & 0.99\leqslant x<1.39\\ \frac{14}{15}, & 1.39\leqslant x<1.65\\ 1, & x\geqslant 1.65\end{cases}$$

The empirical CDF is sketched in Figure 6.1.

Problem 6.3 Suppose X follows the exponential distribution with the PDF

$$f(x)=\begin{cases}\lambda e^{-\lambda x}, & x>0\\ 0, & x\leqslant 0\end{cases}.$$

$(X_1, X_2, \cdots, X_n)$ is a sample from the population X, $X_{(1)}, X_{(2)}, \cdots, X_{(n)}$ are ordered statistics. Find the PDF of the range $Y=X_{(n)}-X_{(1)}$.

Solution:

We first derive the joint CDF and joint PDF of the order statistics $(X_{(1)},X_{(n)})$.

If $x<y$,

$$\begin{aligned}F_{X(1),X(n)}(x,y)&=P(X_{(1)}\leqslant x,X_{(n)}\leqslant y)\\&=P(\text{at least one }X_i\leqslant x,\text{all }X_i\leqslant y)\\&=\sum_{k=1}^{n}P(\text{exact }k\ X_i\text{'s satisfying }X_i\leqslant x,\text{all }X_i\leqslant y)\\&=\sum_{k=1}^{n}P(\text{exact }k\ X_i\text{'s}\leqslant x,\text{other }n-k\ X_i\text{'s satifying }x<X_i\leqslant y)\\&=\sum_{k=1}^{n}\mathrm{C}_n^k[F(x)]^k[F(y)-F(x)]^{n-k}\\&=[F(y)]^n-[F(y)-F(x)]^n.\end{aligned}$$

If $x\geqslant y$,

$$\begin{aligned}F_{X(1),X(n)}(x,y)&=P(X_{(1)}\leqslant x,X_{(n)}\leqslant y)\\&=P(X_{(n)}\leqslant y)\\&=[F(y)]^n.\end{aligned}$$

The joint CDF is

$$F_{X(1),X(n)}(x,y)=\begin{cases}[F(y)]^n-[F(y)-F(x)]^n, & x<y\\ [F(y)]^n, & x\geqslant y\end{cases}.$$

Take the partial derivatives with respect to x and y, respectively, and the joint PDF is obtained.

If $x<y$,

$$\begin{aligned}f_{X(1),X(n)}(x,y)&=\frac{\partial^2}{\partial x\partial y}F_{X(1),X(n)}(x,y)\\&=\frac{\partial}{\partial y}\{n[F(y)-F(x)]^{n-1}f(x)\}\\&=n(n-1)[F(y)-F(x)]^{n-2}f(x)f(y).\end{aligned}$$

If $x\geqslant y$, $f_{X(1),X(n)}(x,y)=\dfrac{\partial^2}{\partial x\partial y}F_{X(1),X(n)}(x,y)=0.$

Therefore, the joint PDF of $(X_{(1)},X_{(n)})$ is

$$f_{X(1),X(n)}(x,y)=\begin{cases}n(n-1)[F(y)-F(x)]^{n-2}f(x)f(y), & x<y\\ 0, & x\geqslant y\end{cases}.$$

Next we derive the density of the range. Let

$$\begin{cases} Y=r_1(X_{(1)},X_{(n)})=X_{(n)}-X_{(1)} \\ Z=r_2(X_{(1)},X_{(n)})=X_{(1)} \end{cases}$$

$$S=\{(x_{(1)},x_{(n)})\mid x_{(1)}<x_{(n)}\}$$

r_1,r_2 define a one-to-one differentiable transformation of S onto a subset T of $\mathbb{R}^2$, as shown in Figure 6.2. The inverse functions can be found as

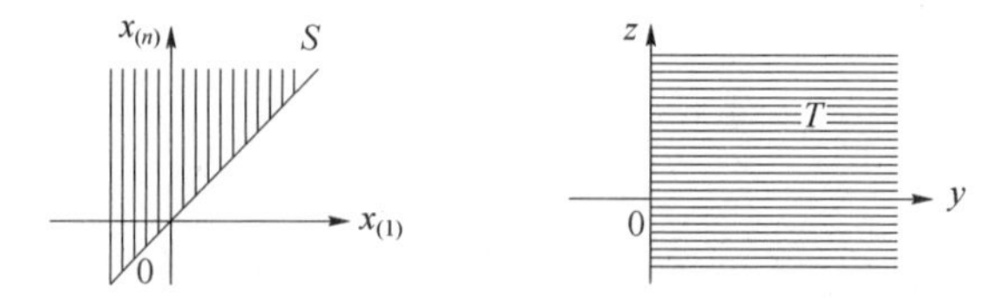

Figure 6.2

$$\begin{cases} X_{(1)}=s_1(Y,Z)=Z \\ X_{(n)}=s_2(Y,Z)=Y+Z \end{cases}$$

Therefore, $Y>0$, $-\infty<Z<+\infty$. And

$$T=\{(y,z)\mid y>0,-\infty<z<+\infty\}$$

$$J=\begin{vmatrix} 0, & 1 \\ 1, & 1 \end{vmatrix}=-1,\ |J|=1$$

Therefore,

$$g(y,z)=\begin{cases} f(s_1,s_2)\times|J|, & (y,z)\in T \\ 0, & \text{otherwise} \end{cases}$$

$$=\begin{cases} n(n-1)[F(y+z)-F(z)]^{n-2}f(z)f(y+z), & (y,z)\in T \\ 0, & \text{otherwise} \end{cases}.$$

Find the marginal PDF $g_Y(y)$ of $Y=X_{(n)}-X_{(1)}$.

(1) If $y\leqslant 0$, $g_Y(y)=0$.

(2) If $y>0$,

$$g_Y(y)=\int_{-\infty}^{+\infty} n(n-1)[F(y+z)-F(z)]^{n-2}f(z)f(y+z)\,\mathrm{d}z$$

$$=\int_{-\infty}^{+\infty} n(n-1)[F(y+x)-F(x)]^{n-2}f(x)f(y+x)\,\mathrm{d}x.$$

Thus, the PDF of $Y=X_{(n)}-X_{(1)}$ is

$$g_Y(y)=\begin{cases} \int_{-\infty}^{+\infty} n(n-1)[F(x+y)-F(x)]^{n-2}f(x)f(x+y)\,\mathrm{d}x, & y>0 \\ 0, & y\leqslant 0 \end{cases}.$$

Now we return to the exponential distribution,

$$f(x)=\begin{cases} \lambda \mathrm{e}^{-\lambda x}, & x>0 \\ 0, & x\leqslant 0 \end{cases},\ F(x)=\begin{cases} 1-\mathrm{e}^{-\lambda x}, & x>0 \\ 0, & x\leqslant 0 \end{cases}.$$

(1) If $y\leqslant 0$, $g_Y(y)=0$.

(2) If $y>0$,

$$g_Y(y)=\int_0^{+\infty} n(n-1)\left[e^{-\lambda x}-e^{-\lambda(x+y)}\right]^{n-2}\lambda^2 e^{-\lambda(2x+y)}dx$$

$$=n(n-1)\lambda^2\left[1-e^{-\lambda y}\right]^{n-2}e^{-\lambda y}\int_0^{+\infty}e^{-n\lambda x}dx$$

$$=(n-1)\lambda\left[1-e^{-\lambda y}\right]^{n-2}e^{-\lambda y}.$$

The desired PDF of $Y=X_{(n)}-X_{(1)}$ is

$$g_Y(y)=\begin{cases}(n-1)\lambda\left[1-e^{-\lambda y}\right]^{n-2}e^{-\lambda y}, & y>0\\ 0, & y\leqslant 0\end{cases}.$$

Problem 6.4 Make a histogram for the following ages of 35 persons.

67	74	60	90	64	72	83
77	67	85	56	57	73	66
60	80	63	90	78	70	63
79	49	88	68	56	78	71
71	46	53	67	64	65	58

**Solution**:

The histogram is plotted in Figure 6.3.

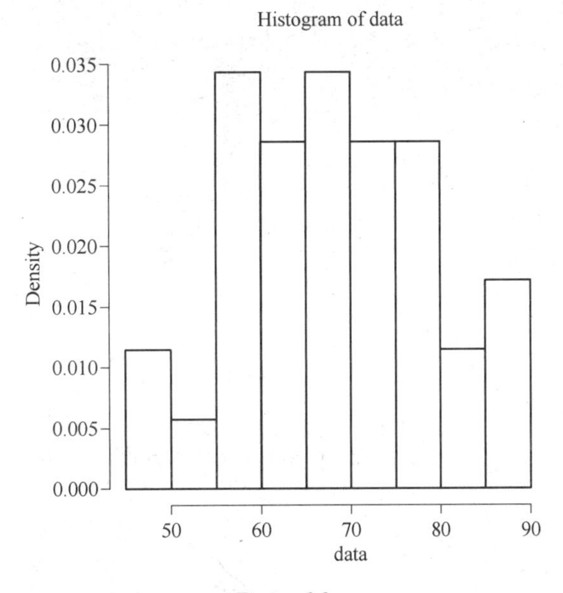

Figure 6.3

Problem 6.5 Suppose $(X_1,\cdots,X_n)$ is a sample from the following population X, determine the moment estimators of the parameters.

(1)

$$X\sim f(x)=\begin{cases}(\theta+1)x^{\theta}, & 0<x<1\\ 0, & \text{otherwise}\end{cases},$$

where $\theta>-1$ is an unknown parameter.

(2)

$$X\sim f(x)=\begin{cases}\dfrac{x}{\theta^2}e^{-\frac{x^2}{2\theta^2}}, & x>0\\ 0, & \text{otherwise}\end{cases},$$

where $\theta>0$ is an unknown parameter.

(3)

$$X\sim f(x)=\begin{cases}\dfrac{1}{\theta}e^{-\frac{x-\mu}{\theta}}, & x\geqslant\mu\\ 0, & \text{otherwise}\end{cases},$$

where $\theta>0$ and μ are unknown parameters.

(4)

$$X\sim f(x)=\begin{cases}1, & \theta-\dfrac{1}{2}\leqslant x\leqslant\theta+\dfrac{1}{2}\\ 0, & \text{otherwise}\end{cases},$$

where θ is an unknown parameter.

Solution:

(1) Since $E(X)=\int_0^1 x(\theta+1)x^{\theta}dx=\dfrac{\theta+1}{\theta+2}$, Let $\dfrac{\theta+1}{\theta+2}=\overline{X}$, we obtain the moment estimator $\hat{\theta}=\dfrac{1-2\overline{X}}{\overline{X}-1}$.

(2) Since $E(X)=\int_0^{+\infty}x\dfrac{x}{\theta^2}e^{-\frac{x^2}{2\theta^2}}dx=\sqrt{2}\theta\int_0^{+\infty}t^{\frac{1}{2}}e^{-t}dt=\sqrt{2}\theta\Gamma\left(\dfrac{3}{2}\right)=\dfrac{\sqrt{2\pi}\theta}{2}$, let $\dfrac{\sqrt{2\pi}\theta}{2}=\overline{X}$, we obtain the moment estimator $\hat{\theta}=\sqrt{\dfrac{2}{\pi}}\overline{X}$.

(3) Since

$$E(X)=\int_{\mu}^{+\infty}x\frac{1}{\theta}e^{-\frac{x-\mu}{\theta}}dx=\int_0^{+\infty}(\mu+\theta t)e^{-t}dt=\mu+\theta,$$

$$E(X^2)=\int_{\mu}^{+\infty}x^2\frac{1}{\theta}e^{-\frac{x-\mu}{\theta}}dx=\int_0^{+\infty}(\mu+\theta t)^2e^{-t}dt=\mu^2+2\mu\theta+2\theta^2.$$

Let $\mu + \theta = \overline{X}, \mu^2 + 2\mu\theta + 2\theta^2 = \frac{1}{n}\sum_{i=1}^{n} X_i^2$.

Solve these equations and obtain the moment estimators

$$\hat{\theta} = \sqrt{\frac{1}{n}\sum_{i=1}^{n}(X_i - \overline{X})^2}, \hat{\mu} = \overline{X} - \sqrt{\frac{1}{n}\sum_{i=1}^{n}(X_i - \overline{X})^2}.$$

(4) Since $E(X) = \int_{\theta-\frac{1}{2}}^{\theta+\frac{1}{2}} x\mathrm{d}x = \theta$, let $\theta = \overline{X}$, we obtain the moment estimator $\hat{\theta} = \overline{X}$.

Problem 6.6 Determine the MLEs of the parameters in the previous problem.

Solution:

(1) Let $x_1, \cdots, x_n$ be the sample observations. Then the likelihood function is

$$L(\theta) = \prod_{i=1}^{n} f(x_i) = (\theta + 1)^n (x_1 x_2 \cdots x_n)^\theta, 0 < x_1, \cdots, x_n < 1.$$

$$\ln L(\theta) = n\ln(\theta + 1) + \theta\sum_{i=1}^{n} \ln x_i.$$

$$\frac{\mathrm{d}\ln L(\theta)}{\mathrm{d}\theta} = \frac{n}{\theta + 1} + \sum_{i=1}^{n} \ln x_i = 0.$$

Thus, $\hat{\theta} = -\frac{n}{\sum_{i=1}^{n} \ln(x_i)} - 1$, so the maximum likelihood estimator of θ is $\hat{\theta} = -\frac{n}{\sum_{i=1}^{n} \ln(X_i)} - 1$.

(2) Let $x_1, \cdots, x_n$ be the sample observations. Then the likelihood function is

$$L(\theta) = \prod_{i=1}^{n} f(x_i) = \frac{x_1 x_2 \cdots x_n}{\theta^{2n}} \mathrm{e}^{-\frac{1}{2\theta^2}\sum_{i=1}^{n} x_i^2}, x_1, \cdots, x_n > 0.$$

$$\ln L(\theta) = \sum_{i=1}^{n} \ln x_i - 2n\ln\theta - \frac{1}{2\theta^2}\sum_{i=1}^{n} x_i^2.$$

$$\frac{\mathrm{d}\ln L(\theta)}{\mathrm{d}\theta} = -\frac{2n}{\theta} + \frac{1}{\theta^3}\sum_{i=1}^{n} x_i^2 = 0.$$

Thus, $\hat{\theta} = \sqrt{\frac{\sum_{i=1}^{n} x_i^2}{2n}}$, so the maximum likelihood estimator of θ is $\hat{\theta} = \sqrt{\frac{\sum_{i=1}^{n} X_i^2}{2n}}$.

(3) Let $x_1, \cdots, x_n$ be the sample observations. Then the likelihood function is

$$L(\theta, \mu) = \prod_{i=1}^{n} f(x_i) = \frac{1}{\theta^n} \mathrm{e}^{-\frac{1}{\theta}\sum_{i=1}^{n}(x_i - \mu)}, x_1, \cdots, x_n \geqslant \mu.$$

$$\ln L(\theta,\mu)=-n\ln\theta-\frac{1}{\theta}\sum_{i=1}^{n}(x_i-\mu),x_1,\cdots,x_n\geqslant\mu.$$

$\frac{\partial\ln L(\theta,\mu)}{\partial\mu}=\frac{n}{\theta}>0$, $L(\theta,\mu)$ is an increasing function of μ, since $\mu\leqslant x_{(1)}=\min\{x_1,\cdots x_n\}$, so $\hat{\mu}=x_{(1)}$.

In addition, let $\frac{\partial\ln L(\theta,\hat{\mu})}{\partial\theta}=-\frac{n}{\theta}+\frac{1}{\theta^2}\sum_{i=1}^{n}(x_i-\hat{\mu})=0$. We can get $\hat{\theta}=\bar{x}-\hat{\mu}=\bar{x}-x_{(1)}$, so the maximum likelihood estimators of θ,μ are $\hat{\theta}=\bar{X}-X_{(1)}$, $\hat{\mu}=X_{(1)}$.

(4) Let $x_1,\cdots,x_n$ be the sample observations. Then the likelihood function is

$$L(\theta)=\prod_{i=1}^{n}f(x_i)=1,\theta-\frac{1}{2}\leqslant x_1,\cdots,x_n\leqslant\theta+\frac{1}{2}.$$

$\theta-\frac{1}{2}\leqslant x_1,\cdots,x_n\leqslant\theta+\frac{1}{2}$ is equivalent to $x_{(n)}-\frac{1}{2}\leqslant\theta\leqslant x_{(1)}+\frac{1}{2}$, where $x_{(n)}=\max\{x_1,\cdots,x_n\}$, $x_{(1)}=\min\{x_1,\cdots,x_n\}$.

Then if θ takes any point on the interval $\left[x_{(n)}-\frac{1}{2},x_{(1)}+\frac{1}{2}\right]$, $L(\theta)$ achieves the maximum value of 1. The maximum likelihood estimator of θ is any point on the interval $\left[X_{(n)}-\frac{1}{2},X_{(1)}+\frac{1}{2}\right]$.

Problem 6.7 Suppose $X\sim B(n,p)$, $0<p<1$, n is known. $(X_1,\cdots,X_n)$ is a sample from the population X, determine the moment estimator and maximum likelihood estimator of the parameter p.

 **Solution**:

(1) Since $E(X)=np$, let $np=\bar{X}$, we obtain the moment estimator $\hat{p}=\frac{\bar{X}}{n}$.

(2) The likelihood function is

$$L(p)=\prod_{i=1}^{n}f(x_i)=\prod_{i=1}^{n}C_n^{x_i}p^{x_i}(1-p)^{n-x_i}=\left(\prod_{i=1}^{n}C_n^{x_i}\right)p^{\sum_{i=1}^{n}x_i}(1-p)^{n^2-\sum_{i=1}^{n}x_i}.$$

The log-likelihood function is

$$\ln L(p)=\ln\left(\prod_{i=1}^{n}C_n^{x_i}\right)+\sum_{i=1}^{n}x_i\ln(p)+\left(n^2-\sum_{i=1}^{n}x_i\right)\ln(1-p).$$

Let $\frac{\mathrm{d}\ln(p)}{\mathrm{d}p}=\frac{1}{p}\sum_{i=1}^{n}x_i-\frac{1}{1-p}\left(n^2-\sum_{i=1}^{n}x_i\right)=0$. Solve the equation, we have $\hat{p}_{\mathrm{MLE}}=\frac{\bar{x}}{n}$. The maximum likelihood estimator is $\frac{\bar{X}}{n}$.

Problem 6.8 Suppose $X \sim P(\lambda)$, where $\lambda > 0$ is a parameter. $(X_1, \cdots, X_n)$ is a sample from the population X.

(1) Find the maximum likelihood estimator of λ.

(2) Is the maximum likelihood estimator unbiased?

Solution:

(1) The likelihood function is

$$L(\lambda) = \prod_{i=1}^{n} f(x_i) = \prod_{i=1}^{n} \frac{\lambda^{x_i}}{x_i!} e^{-\lambda} = \frac{\lambda^{\sum_{i=1}^{n} x_i} e^{-n\lambda}}{x_1! x_2! \cdots x_n!}.$$

The log-likelihood function is

$$\ln L(\lambda) = \left(\sum_{i=1}^{n} x_i\right) \ln\lambda - n\lambda - \ln(x_1! x_2! \cdots x_n!).$$

Let $\frac{\mathrm{d}\ln L(\lambda)}{\mathrm{d}\lambda} = \frac{1}{\lambda} \sum_{i=1}^{n} x_i - n = 0$, we can get $\hat{\lambda} = \bar{x}$, so the maximum likelihood estimator of λ is $\hat{\lambda} = \bar{X}$.

(2) $E(\bar{X}) = E\left(\frac{X_1 + X_2 + \cdots X_n}{n}\right) = E(X) = \lambda$. The maximum likelihood estimator is unbiased.

Problem 6.9 Suppose that $X_1, \cdots, X_n$ form a random sample from the uniform distribution on the interval $[\theta_1, \theta_2]$, where both θ_1 and θ_2 are unknown $(-\infty < \theta_1 < \theta_2 < +\infty)$. Find the maximum likelihood estimators of θ_1 and θ_2.

Solution:

The PDF of $X_i (i = 1, 2, \cdots, n)$ is as follows:

$$f(x \mid \theta_1, \theta_2) = \begin{cases} \frac{1}{\theta_2 - \theta_1}, & \theta_1 \leqslant x \leqslant \theta_2 \\ 0, & \text{otherwise} \end{cases}.$$

Therefore, the likelihood function is

$$L(\theta_1, \theta_2) = \frac{1}{(\theta_2 - \theta_1)^n}$$

for $\theta_1 \leqslant \min\{x_1, \cdots, x_n\} < \max\{x_1, \cdots, x_n\} \leqslant \theta_2$, and $L(\theta_1, \theta_2) = 0$, otherwise.

Hence, $L(\theta_1, \theta_2)$ acheives its maximum when $\theta_2 - \theta_1$ takes the smallest value. Since the smallest value of θ_2 is $\max\{x_1, \cdots, x_n\}$ and the largest value of θ_1 is $\min\{x_1, \cdots, x_n\}$, thus, the MLEs are

$$\hat{\theta}_2 = \max\{X_1, \cdots, X_n\}, \hat{\theta}_1 = \min\{X_1, \cdots, X_n\}.$$

Problem 6.10 Suppose that $X_1, \cdots, X_n$ form a random sample from an exponential distribution for which the value of the parameter λ is unknown. Determine the maximum likelihood estimator of the median of the distribution.

Solution:

The PDF of the exponential distribution $\mathrm{E}(\lambda)$ is

$$f(x)=\begin{cases}\lambda\, \mathrm{e}^{-\lambda x}, & x\geqslant 0\\ 0, & \text{otherwise}\end{cases}.$$

The median of the distribution is the number m satisfying

$$\int_0^m f(x)\,\mathrm{d}x = \int_0^m \lambda\, \mathrm{e}^{-\lambda x}\,\mathrm{d}x = \frac{1}{2}.$$

Therefore, $m=\dfrac{\ln 2}{\lambda}$.

Let $x_1, \cdots, x_n$ be the sample observations. Then the likelihood function is

$$L(\lambda) = \prod_{i=1}^n f(x_i) = \prod_{i=1}^n \lambda\, \mathrm{e}^{-\lambda x_i} = \lambda^n\, \mathrm{e}^{-\lambda \sum\limits_{i=1}^n x_i}.$$

The log-likelihood function is

$$\ln L(\lambda) = n\ln\lambda - \lambda \sum_{i=1}^n x_i.$$

Let $\dfrac{\mathrm{d}\ln L(\lambda)}{\mathrm{d}\lambda} = \dfrac{n}{\lambda} - \sum\limits_{i=1}^n x_i = 0$, we can get $\hat{\lambda}=\dfrac{1}{\bar{x}}$, so the maximum likelihood estimator of λ is $\hat{\lambda}=\dfrac{1}{\bar{X}}$. By the invariance property of MLE,

$$\hat{m}=\frac{\ln 2}{\hat{\lambda}}=(\ln 2)\bar{X}.$$

Problem 6.11 Suppose that $X_1, \cdots, X_n$ form a random sample from a normal distribution X for which both the mean and the variance are unknown. Find the maximum likelihood estimator of the 0.95 quantile of the distribution, that is, of the point x_0 such that $P(X\leqslant x_0)=0.95$.

Solution:

$X\sim \mathrm{N}(\mu,\sigma^2)$. $Z=\dfrac{X-\mu}{\sigma}\sim \mathrm{N}(0,1)$. Therefore,

$$0.95=P(X\leqslant x_0)=P\left(Z<\frac{x_0-\mu}{\sigma}\right)=\Phi\left(\frac{x_0-\mu}{\sigma}\right).$$

From the standard normal table, $\dfrac{x_0-\mu}{\sigma}=\Phi^{-1}(0.95)=1.645$. Then $x_0=\mu+1.645\sigma$.

By the invariance property of MLE, it follows that $\hat{x_0}=\hat{\mu}+1.645\,\hat{\sigma}$, where

$$\hat{\mu} = \overline{X}, \hat{\sigma} = \sqrt{\frac{1}{n}\sum_{i=1}^{n}(X_i - \overline{X})^2}.$$

Problem 6.12 The PDF of X is given by

$$f(x)=\begin{cases}\frac{1}{\theta}, & 0<x\leqslant\theta\\ 0, & \text{otherwise}\end{cases},$$

where $\theta>0$ is an unknown parameter. (X_1, X_2, X_3) is a sample from X. Show that $\frac{4}{3}\max\limits_{1\leqslant i\leqslant 3}\{X_i\}$ and $4\min\limits_{1\leqslant i\leqslant 3}\{X_i\}$ are unbiased estimators of θ. Which one is more efficient?

 **Solution**:

The PDF of X is given by

$$f(x)=\begin{cases}\frac{1}{\theta}, & 0<x\leqslant\theta\\ 0, & \text{otherwise}\end{cases}.$$

The CDF of X is given by

$$F(x)=\begin{cases}0, & x\leqslant 0\\ \frac{x}{\theta}, & 0<x\leqslant\theta.\\ 1, & x>\theta\end{cases}$$

The PDF of $X_{(3)}=\max\{X_1, X_2, X_3\}$ is

$$f_{X_{(3)}}(x)=3[F(x)]^2 f(x)=\begin{cases}\frac{3x^2}{\theta^3}, & 0<x\leqslant\theta\\ 0, & \text{otherwise}\end{cases}.$$

$$E\left(\frac{4}{3}\max_{1\leqslant i\leqslant 3}\{X_i\}\right)=\frac{4}{3}\int_0^{\theta} x\times\frac{3x^2}{\theta^3}dx=\theta,$$

so $\frac{4}{3}\max\limits_{1\leqslant i\leqslant 3}\{X_i\}$ is an unbiased estimator of θ.

$$E\left[\left(\frac{4}{3}\max_{1\leqslant i\leqslant 3}\{X_i\}\right)^2\right]=\left(\frac{4}{3}\right)^2\int_0^{\theta} x^2\times\frac{3x^2}{\theta^3}dx=\frac{16}{15}\theta^2.$$

$$V\left(\frac{4}{3}\max_{1\leqslant i\leqslant 3}\{X_i\}\right)=\frac{16}{15}\theta^2-\theta^2=\frac{1}{15}\theta^2.$$

The PDF of $X_{(1)}=\min\{X_1, X_2, X_3\}$ is

$$f_{X_{(1)}}(x)=3[1-F(x)]^2 f(x)=\begin{cases}\frac{3}{\theta}\left(1-\frac{x}{\theta}\right)^2, & 0<x\leqslant\theta\\ 0, & \text{otherwise}\end{cases}.$$

$$E\left(4\min_{1\leqslant i\leqslant 3}\{X_i\}\right)=4\int_0^\theta x\times\frac{3}{\theta}\left(1-\frac{x}{\theta}\right)^2\mathrm{d}x=12\theta\int_0^1 t(1-t)^2\mathrm{d}t=\theta,$$

so $4\min_{i\leqslant i\leqslant 3}\{X_i\}$ is an unbiased estimator of θ.

$$E[(4\min_{1\leqslant i\leqslant 3}\{X_i\})^2]=16\int_0^\theta x^2\times\frac{3}{\theta}\left(1-\frac{x}{\theta}\right)^2\mathrm{d}x=48\,\theta^2\int_0^1 t(1-t)^2\mathrm{d}t=\frac{8}{5}\theta^2.$$

$$V\left(4\min_{1\leqslant i\leqslant 3}\{X_i\}\right)=\frac{8}{5}\theta^2-\theta^2=\frac{3}{5}\theta^2.$$

$V\left(\frac{4}{3}\max_{1\leqslant i\leqslant 3}\{X_i\}\right)<V\left(4\min_{1\leqslant i\leqslant 3}\{X_i\}\right)$, therefore, $\frac{4}{3}\max_{1\leqslant i\leqslant 3}\{X_i\}$ is more efficient.

Problem 6.13 $(X_1,\cdots,X_n)$ is a sample from X, determine the value of k such that $\hat{\sigma^2}=k\sum_{i=1}^{n-1}(X_{i+1}-X_i)^2$ is an unbiased estimator of the population variance σ^2.

Solution:

$$\begin{aligned}E(\hat{\sigma^2})&=E\left[k\sum_{i=1}^{n-1}(X_{i+1}-X_i)^2\right]\\&=k\sum_{i=1}^{n-1}E(X_{i+1}^2-2X_{i+1}X_i+X_i^2)\\&=k\sum_{i=1}^{n-1}[E(X_{i+1}^2)-2E(X_{i+1})E(X_i)+E(X_i^2)]\\&=k\sum_{i=1}^{n-1}[2E(X^2)-2[E(X)]^2]\\&=2k(n-1)\sigma^2.\end{aligned}$$

Let $2k(n-1)\sigma^2=\sigma^2$, we obtain $k=\frac{1}{2(n-1)}$.

Problem 6.14 Suppose $X\sim N(\mu,\sigma^2)$, (X_1,X_2,X_3) is a sample from population X. Let

$$\hat{\mu}_1=\frac{1}{5}X_1+\frac{3}{10}X_2+\frac{1}{2}X_3,$$

$$\hat{\mu}_2=\frac{1}{3}X_1+\frac{1}{4}X_2+\frac{5}{12}X_3,$$

$$\hat{\mu}_3=\frac{1}{3}X_1+\frac{1}{6}X_2+\frac{1}{2}X_3.$$

Show that the three statistics are unbiased estimators of μ. Which one is the most efficient?

Solution:

$$E(\hat{\mu}_1)=\frac{1}{5}E(X_1)+\frac{3}{10}E(X_2)+\frac{1}{2}E(X_3)=\mu,$$

$$E(\hat{\mu}_2)=\frac{1}{3}E(X_1)+\frac{1}{4}E(X_2)+\frac{5}{12}E(X_3)=\mu,$$

$$E(\hat{\mu}_3)=\frac{1}{3}E(X_1)+\frac{1}{6}E(X_2)+\frac{1}{2}E(X_3)=\mu,$$

so the three statistics are unbiased estimators of μ.

$$V(\hat{\mu}_1)=\left(\frac{1}{5}\right)^2V(X_1)+\left(\frac{3}{10}\right)^2V(X_2)+\left(\frac{1}{2}\right)^2V(X_3)=\frac{19}{50}\sigma^2.$$

$$V(\hat{\mu}_2)=\left(\frac{1}{3}\right)^2V(X_1)+\left(\frac{1}{4}\right)^2V(X_2)+\left(\frac{5}{12}\right)^2V(X_3)=\frac{25}{72}\sigma^2.$$

$$V(\hat{\mu}_3)=\left(\frac{1}{3}\right)^2V(X_1)+\left(\frac{1}{6}\right)^2V(X_2)+\left(\frac{1}{2}\right)^2V(X_3)=\frac{7}{18}\sigma^2.$$

Since $\frac{25}{72}\sigma^2<\frac{19}{50}\sigma^2<\frac{7}{18}\sigma^2$, $\hat{\mu}_2$ is the most efficient.

Problem 6.15 Suppose that X is a random variable whose distribution is completely unknown, but it is known that all the moments $E(X^k)$ $(k=1,2,\cdots)$ are finite. Suppose also that $X_1,\cdots,X_n$ form a random sample from this distribution. Show that the k-th sample moment $\frac{1}{n}\sum_{i=1}^{n}X^k$ is an unbiased estimator of $E(X^k)$.

Solution:

$$E\left(\frac{1}{n}\sum_{i=1}^{n}X_i^k\right)=\frac{1}{n}\sum_{i=1}^{n}E(X_i^k)=\frac{1}{n}\sum_{i=1}^{n}E(X^k)=E(X^k).$$

Problem 6.16 Suppose that $X_1,\cdots,X_n$ form a random sample from the geometric distribution with unknown parameter p. Find a statistic that will be an unbiased estimator of $1/p$.

Solution:

If X has the geometric distribution with parameter p, then $E(X)=1/p$. On the other hand, $E(\overline{X})=E(X)$, which implies that $\overline{X}$ is a statistic that is an unbiased estimator of $1/p$.

Problem 6.17 Suppose that $X_1,\cdots,X_n$ form a random sample from the uniform distribution on the interval $[0,\theta]$, where the value of the parameter θ is unknown; and let $Y_n=\max\{X_1,\cdots,X_n\}$. Show that $[(n+1)/n]Y_n$ is an unbiased estimator of θ.

Solution:

$X\sim U[0,\theta]$. The PDF of X is

$$f(x)=\begin{cases}\dfrac{1}{\theta}, & 0\leqslant x\leqslant\theta \\ 0, & \text{otherwise}\end{cases}.$$

The CDF of X is given by

$$F(x)=\begin{cases}0, & x<0 \\ \dfrac{x}{\theta}, & 0\leqslant x\leqslant\theta. \\ 1, & x>\theta\end{cases}$$

The PDF of $Y_n=\max\{X_1,\cdots,X_n\}$ is

$$f_{Y_n}(x)=n[F(x)]^{n-1}f(x)=\begin{cases}\dfrac{n\,x^{n-1}}{\theta^n}, & 0\leqslant x\leqslant\theta \\ 0, & \text{otherwise}\end{cases}.$$

$$E\left(\frac{n+1}{n}Y_n\right)=\frac{n+1}{n}\int_0^\theta x\times\frac{n\,x^{n-1}}{\theta^n}\mathrm{d}x=\theta.$$

Therefore, $\dfrac{n+1}{n}Y_n$ is an unbiased estimator of θ.

Problem 6.18 Suppose $X_1,\cdots,X_{36}$ form a random sample from $X\sim N(55,6.3^2)$. Find the probability that the sample mean falls into the interval $(53.8,56.8)$.

 **Solution**:

$\overline{X}\sim N\left(55,\dfrac{6.3^2}{36}\right)=N(55,1.05^2).Z=\dfrac{\overline{X}-55}{1.05}\sim N(0,1).$

$$\begin{aligned}P(53.8<\overline{X}<56.8)&=P\left(\frac{53.8-55}{1.05}<\frac{\overline{X}-55}{1.05}<\frac{56.8-55}{1.05}\right)\\&=P(-1.14<Z<1.71)\\&=\Phi(1.71)+\Phi(1.14)-1\\&=0.829\,3.\end{aligned}$$

Problem 6.19 Suppose $X\sim N(0,1)$, $(X_1,X_2,\cdots,X_n)$ forms a random sample from X. Find the distributions of the following statistics:

(1) X_1-X_2;

(2) $\dfrac{X_1-X_2}{\sqrt{X_3^2+X_4^2}}$;

(3) $\dfrac{\sqrt{n-1}\,X_n}{\sqrt{\sum\limits_{i=1}^{n-1}X_i^2}}$;

(4) $\frac{n-5}{5} \times \frac{\sum_{i=1}^{5} X_i^2}{\sum_{i=6}^{n} X_i^2}$.

Solution:

(1) $X_1 \sim N(0,1)$, $X_2 \sim N(0,1)$, X_1 and X_2 are independent, thus, $X_1 - X_2 \sim N(0,2)$.

(2) $X_1 - X_2 \sim N(0,2)$, so $\frac{X_1 - X_2}{\sqrt{2}} \sim N(0,1)$. $X_3 \sim N(0,1)$, so $X_3^2 \sim \chi^2(1)$. Similarly, $X_4^2 \sim \chi^2(1)$. Therefore, $X_3^2 + X_4^2 \sim \chi^2(2)$.

$$\frac{X_1 - X_2}{\sqrt{X_3^2 + X_4^2}} = \frac{\frac{X_1 - X_2}{\sqrt{2}}}{\sqrt{\frac{X_3^2 + X_4^2}{2}}} \sim t(2).$$

(3) $X_n \sim N(0,1)$, $\sum_{i=1}^{n-1} X_i^2 \sim \chi^2(n-1)$. Therefore,

$$\frac{\sqrt{n-1}\, X_n}{\sqrt{\sum_{i=1}^{n-1} X_i^2}} = \frac{X_n}{\sqrt{\frac{\sum_{i=1}^{n-1} X_i^2}{n-1}}} \sim t(n-1).$$

(4) $\sum_{i=1}^{5} X_i^2 \sim \chi^2(5)$, $\sum_{i=6}^{n} X_i^2 \sim \chi^2(n-5)$. Therefore,

$$\frac{n-5}{5} \times \frac{\sum_{i=1}^{5} X_i^2}{\sum_{i=6}^{n} X_i^2} = \frac{\frac{\sum_{i=1}^{5} X_i^2}{5}}{\frac{\sum_{i=6}^{n} X_i^2}{n-5}} \sim F(5, n-5).$$

Problem 6.20 Suppose that $X_1, \cdots, X_n$ form a random sample from the normal distribution with mean μ and variance σ^2. Find the distribution of

$$\frac{n(\overline{X} - \mu)^2}{\sigma^2}.$$

Solution:

It is known that $\overline{X}$ has the normal distribution with mean μ and variance $\frac{\sigma^2}{n}$. Therefore, $\frac{\overline{X} - \mu}{\frac{\sigma}{\sqrt{n}}}$

has the standard normal distribution and the square of this variable has the χ^2-distribution with one degree of freedom.

$$\frac{n(\overline{X}-\mu)^2}{\sigma^2}\sim\chi^2(1).$$

Problem 6.21 Suppose that six random variables $X_1,\cdots,X_6$ form a random sample from the standard normal distribution, and let

$$Y=(X_1+X_2+X_3)^2+(X_4+X_5+X_6)^2.$$

Determine a value of c such that the random variable cY will have a χ^2-distribution.

Solution:

Since $X_1,\cdots,X_6$ form a random sample from the standard normal distribution, $X_1+X_2+X_3\sim N(0,3)$, $X_4+X_5+X_6\sim N(0,3)$. Thus, $(X_1+X_2+X_3)/\sqrt{3}\sim N(0,1)$, $(X_4+X_5+X_6)/\sqrt{3}\sim N(0,1)$. Therefore,

$$\frac{1}{3}Y=\left(\frac{X_1+X_2+X_3}{\sqrt{3}}\right)^2+\left(\frac{X_4+X_5+X_6}{\sqrt{3}}\right)^2\sim\chi^2(2).$$

We can choose $c=\frac{1}{3}$, such that the random variable cY will have a χ^2-distribution.

Problem 6.22 Suppose that the random variables $X_1,\cdots,X_5$ are IID and that each has the standard normal distribution. Determine a constant c such that the random variable $\frac{c(X_1+X_2)}{(X_3^2+X_4^2+X_5^2)^{\frac{1}{2}}}$ will have a t-distribution.

Solution:

X_1+X_2 has the normal distribution with mean 0 and variance 2. Therefore, $Y=\frac{X_1+X_2}{\sqrt{2}}$ has the standard normal distribution. Also, $W=X_3^2+X_4^2+X_5^2$ has the χ^2-distribution with 3 degrees of freedom, and Y and W are independent. Therefore, $\frac{Y}{\sqrt{\frac{W}{3}}}=\frac{\sqrt{\frac{3}{2}}(X_1+X_2)}{(X_3^2+X_4^2+X_5^2)^{\frac{1}{2}}}$ has the t-distribution with 3 degrees of freedom. Thus, if we choose $c=\sqrt{\frac{3}{2}}$, the given random variable will follow t(3).

Problem 6.23 Suppose that the random variables X_1 and X_2 are independent and that each has the normal distribution with mean $\mu = 0$ and variance σ^2. Determine the value of $P\left[\frac{(X_1+X_2)^2}{(X_1-X_2)^2}<4\right]$.

Solution 1:

$X_1 \sim N(0,\sigma^2)$, $X_2 \sim N(0,\sigma^2)$, X_1 and X_2 are independent, the joint PDF of (X_1,X_2) is

$$f(x_1,x_2)=f(x_1)f(x_2)=\frac{1}{2\pi\sigma^2}e^{-\frac{x_1^2+x_2^2}{2\sigma^2}}.$$

$$\begin{cases} U=r_1(X_1,X_2)=\dfrac{X_1+X_2}{X_1-X_2} \\ V=r_2(X_1,X_2)=X_1-X_2 \end{cases}$$

$$S=\{(x_1,x_2)\mid -\infty<x_1,x_2<+\infty\}=\mathbb{R}^2$$

r_1,r_2 define a one-to-one differentiable transformation of S onto a subset T of $\mathbb{R}^2$. The inverse functions can be found as

$$\begin{cases} X_1=s_1(U,V)=\dfrac{(U+1)V}{2} \\ X_2=s_2(U,V)=\dfrac{(U-1)V}{2} \end{cases}.$$

Therefore, $-\infty<U,V<+\infty$. And $T=\{(u,v)\mid -\infty<u,v<+\infty\}=\mathbb{R}^2$.

$$J=\begin{vmatrix} \dfrac{\partial s_1}{\partial u} & \dfrac{\partial s_1}{\partial v} \\ \dfrac{\partial s_2}{\partial u} & \dfrac{\partial s_2}{\partial v} \end{vmatrix}=\begin{vmatrix} \dfrac{v}{2} & \dfrac{u+1}{2} \\ \dfrac{v}{2} & \dfrac{u-1}{2} \end{vmatrix}=-\frac{v}{2},\ |J|=\frac{|v|}{2}.$$

Therefore, the joint PDF of (U,V) is

$$\begin{aligned} g(u,v)&=f(s_1,s_2)\times|J| \\ &=\frac{1}{2\pi\sigma^2}e^{-\frac{(u^2+1)v^2}{4\sigma^2}}\times\frac{|v|}{2},(u,v)\in T=\mathbb{R}^2. \end{aligned}$$

- Find the marginal PDF $g_U(u)$ of U.

$$\begin{aligned} g_U(u)&=\int_{-\infty}^{\infty}g(u,v)\,dv \\ &=\int_{-\infty}^{\infty}\frac{1}{2\pi\sigma^2}e^{-\frac{(u^2+1)v^2}{4\sigma^2}}\times\frac{|v|}{2}dv \\ &=\int_0^{\infty}\frac{1}{2\pi\sigma^2}e^{-\frac{(u^2+1)v^2}{4\sigma^2}}\times v\,dv\ \left(\text{let } t=\frac{(u^2+1)v^2}{4\sigma^2}\right) \end{aligned}$$

$$= \frac{1}{\pi(u^2+1)} \int_0^{\infty} e^{-t} dt$$

$$= \frac{1}{\pi(u^2+1)}.$$

In summary,

$$g_U(u) = \frac{1}{\pi(u^2+1)}, -\infty<u<+\infty.$$

U follows the Cauchy distribution. Therefore,

$$P\left[\frac{(X_1+X_2)^2}{(X_1-X_2)^2}<4\right] = P(U^2<4)$$

$$= P(-2<U<2)$$

$$= \int_{-2}^{2} \frac{1}{\pi(u^2+1)} du$$

$$= \frac{1}{\pi} \arctan u \Big|_{u=-2}^{u=2}$$

$$= \frac{2}{\pi} \arctan 2$$

$$\approx 0.704\ 8.$$

 **Solution 2:**

$X_1 \sim N(0,\sigma^2)$, $X_2 \sim N(0,\sigma^2)$, X_1 and X_2 are independent, $(X_1, X_2) \sim N(0,0,\sigma^2,\sigma^2,0)$. Define

$$\begin{cases} U = X_1 + X_2 \\ V = X_1 - X_2 \end{cases},$$

$$\begin{aligned} \mathrm{Cov}(U,V) &= \mathrm{Cov}(X_1+X_2, X_1-X_2) \\ &= V(X_1) - \mathrm{Cov}\ (X_1,X_2) + \mathrm{Cov}\ (X_2,X_1) - V(X_2) \\ &= \sigma^2 - 0 + 0 - \sigma^2 \\ &= 0. \end{aligned}$$

That is, U, V are uncorrelated. On the other hand, $\begin{bmatrix} U \\ V \end{bmatrix} = \begin{bmatrix} 1 & 1 \\ 1 & -1 \end{bmatrix} \begin{bmatrix} X_1 \\ X_2 \end{bmatrix}$. The matrix $\boldsymbol{A} = \begin{bmatrix} 1 & 1 \\ 1 & -1 \end{bmatrix}$ is full row rank, $|\boldsymbol{A}| \neq 0$, rank$(\boldsymbol{A}) = 2$, therefore, (U, V) follows two dimensional non-degenerate normal distribution.

U and V are uncorrelated if and only if U and V are independent, so $U = X_1 + X_2$ and $V = X_1 - X_2$ are independent.

$$U \sim N(0, 2\sigma^2), \frac{U}{\sqrt{2}\sigma} \sim N(0,1)$$

$$V \sim N(0,2\sigma^2),\frac{V}{\sqrt{2}\sigma} \sim N(0,1),\frac{V^2}{2\sigma^2} \sim \chi^2(1)$$

Therefore,

$$T=\frac{\frac{U}{\sqrt{2}\sigma}}{\sqrt{\frac{V^2}{2\sigma^2}}}=\frac{U}{V} \sim t(1).$$

t(1) is also known as the Cauchy distribution. The following steps are the same as solution 1.

$$\begin{aligned} P\left[\frac{(X_1+X_2)^2}{(X_1-X_2)^2}<4\right] &= P(T^2<4) \\ &= P(-2<T<2) \\ &= \int_{-2}^{2} \frac{1}{\pi(t^2+1)} dt \\ &= \frac{1}{\pi} \arctan t \Big|_{t=-2}^{t=2} \\ &= \frac{2}{\pi} \arctan 2 \\ &\approx 0.704\,8. \end{aligned}$$

 **Solution 3:**

$X_1 \sim N(0,\sigma^2)$, $X_2 \sim N(0,\sigma^2)$, X_1 and X_2 are independent, (X_1,X_2) can be considered as a sample of size $n=2$ from $N(0,\sigma^2)$.

Sample mean $\bar{X}=\frac{1}{2}(X_1+X_2)$.

Sample variance

$$\begin{aligned} S^2 &= (X_1-\bar{X})^2+(X_2-\bar{X})^2 \\ &= \frac{(X_1-X_2)^2}{2}. \end{aligned}$$

Therefore,

$$T=\frac{\bar{X}-0}{\frac{S}{\sqrt{n}}}=\frac{X_1+X_2}{\sqrt{(X_1-X_2)^2}} \sim t(1).$$

t(1) is also known as the Cauchy distribution. The following steps are the same as solution 1.

$$\begin{aligned} P\left[\frac{(X_1+X_2)^2}{(X_1-X_2)^2}<4\right] &= P(T^2<4) \\ &= P(-2<T<2) \\ &= \int_{-2}^{2} \frac{1}{\pi(t^2+1)} dt \end{aligned}$$

$$= \frac{1}{\pi}\arctan t \Big|_{t=-2}^{t=2}$$

$$= \frac{2}{\pi}\arctan 2$$

$$\approx 0.704\,8.$$

Problem 6.24 If $X \sim U(0,1)$, show that $-2\ln X \sim \chi^2(2)$.

Solution:

Let $Y=-2\ln X$, we shall determine the PDF $g(y)$ of Y. The PDF of X is

$$f(x)=\begin{cases}1, & 0<x<1 \\ 0, & \text{otherwise}\end{cases}.$$

Since $y=-2\ln x$ is a differentiable and one-to-one function, the inverse function is $x=e^{-\frac{y}{2}}$, and $\frac{dx}{dy}=-\frac{1}{2}e^{-\frac{y}{2}}$. If $0<x<1$, then $y>0$.

Therefore, for $y>0$,

$$g(y)=f(e^{-\frac{y}{2}})\left|\frac{dx}{dy}\right|=\frac{1}{2}e^{-\frac{y}{2}}.$$

If $y\leqslant 0, g(y)=0$. It is the PDF of $\chi^2(2)$. Therefore, $-2\ln X \sim \chi^2(2)$.

Problem 6.25 $(X_1, X_2, \cdots, X_{10})$ is a sample from the population $X \sim N(0, 0.3^2)$, compute $P\left(\sum_{i=1}^{10} X_i^2 > 1.44\right)$.

Solution:

$$X \sim N(0,0.3^2), \frac{X}{0.3} \sim N(0,1), \frac{X^2}{0.09} \sim \chi^2(1).$$

$$P\left(\sum_{i=1}^{10} X_i^2 > 1.44\right) = P\left(\frac{\sum_{i=1}^{10} X_i^2}{0.09} > \frac{1.44}{0.09}\right) = P(\chi^2(10) > 16) = 0.099\,6.$$

Note: the final answer can be calculated using the Chi-square table or computer program.

Problem 6.26 Suppose that a point (X,Y) is to be chosen at random in the xy-plane, where X and Y are independent random variables and each has the standard normal distribution. If a circle is drawn in the xy-plane with its center at the origin, what is the radius of the smallest circle that can be chosen in order for there to be probability 0.99 that the point (X,Y) will lie inside the circle?

Solution:

Let r denote the radius of the circle. The point (X,Y) will lie inside the circle if and only if $X^2+Y^2\leqslant r^2$. Also, X^2+Y^2 has the χ^2-distribution with two degrees of freedom.

$$P(X^2+Y^2\leqslant r^2)=P(\chi^2(2)\leqslant r^2)\geqslant 0.99.$$

Using the Chi-square table, $P(\chi^2(2)\leqslant 9.210)=0.99$. Therefore, $r^2\geqslant 9.210$, $r\geqslant 3.035$.

Problem 6.27 Suppose that $X_1,\cdots,X_n$ form a random sample from the normal distribution with mean μ and variance σ^2. Assuming that the sample size n is 16, determine the values of the following probabilities:

(1) $P\left[\frac{1}{2}\sigma^2\leqslant\frac{1}{n}\sum_{i=1}^{n}(X_i-\mu)^2\leqslant 2\sigma^2\right]$;

(2) $P\left[\frac{1}{2}\sigma^2\leqslant\frac{1}{n}\sum_{i=1}^{n}(X_i-\bar{X})^2\leqslant 2\sigma^2\right]$.

Solution:

(1) $X_i\sim N(\mu,\sigma^2)$, $\frac{X_i-\mu}{\sigma}\sim N(0,1)$, $\frac{(X_i-\mu)^2}{\sigma^2}\sim\chi^2(1)(i=1,2,\cdots,16)$, $\sum_{i=1}^{n}\frac{(X_i-\mu)^2}{\sigma^2}\sim\chi^2(16)$. Therefore,

$$\begin{aligned}&P\left[\frac{1}{2}\sigma^2\leqslant\frac{1}{n}\sum_{i=1}^{n}(X_i-\mu)^2\leqslant 2\sigma^2\right]\\&=P\left[\frac{1}{2}n\leqslant\sum_{i=1}^{n}\frac{(X_i-\mu)^2}{\sigma^2}\leqslant 2n\right]\\&=P[8\leqslant\chi^2(16)\leqslant 32]\\&=0.99-0.051\ 1\\&=0.938\ 9.\end{aligned}$$

In the second to last step, 0.99 and 0.051 1 can be found using the Chi-square table or computer program.

(2)

$$S^2=\frac{1}{n-1}\sum_{i=1}^{n}(X_i-\bar{X})^2,\text{ and }\frac{(n-1)S^2}{\sigma^2}\sim\chi^2(n-1).$$

Therefore,

$$\begin{aligned}&P\left[\frac{1}{2}\sigma^2\leqslant\frac{1}{n}\sum_{i=1}^{n}(X_i-\bar{X})^2\leqslant 2\sigma^2\right]\\&=P\left[\frac{1}{2}n\leqslant\frac{(n-1)S^2}{\sigma^2}\leqslant 2n\right]\\&=P[8\leqslant\chi^2(15)\leqslant 32]\end{aligned}$$

$$= 0.993\,6 - 0.076\,2$$

$$= 0.917\,4.$$

In the second to last step, 0.993 6 and 0.076 2 can be found using the Chi-square table or computer program.

Problem 6.28 Suppose that $X_1, \cdots, X_n$ form a random sample from the normal distribution with mean μ and variance σ^2, and let S^2 denote the sample variance. Determine the smallest values of n for which the following relations are satisfied:

(1) $P\left(\frac{S^2}{\sigma^2} \leqslant 1.5\right) \geqslant 0.95$;

(2) $P\left(|S^2-\sigma^2| \leqslant \frac{1}{2}\sigma^2\right) \geqslant 0.8$.

Solution:

(1) The random variable $V=\frac{(n-1)S^2}{\sigma^2}$ has the χ^2-distribution with $n-1$ degrees of freedom. The required probability can be written in the form $P(V \leqslant 1.5(n-1)) \geqslant 0.95$. By trial and error, using computer program, it is found that for $n=26$, V has 25 degrees of freedom and $P(V \leqslant 37.5) = 0.948\,3 < 0.95$. However, for $n=27$, V has 26 degrees of freedom and $P(V \leqslant 39.0) = 0.951\,2 > 0.95$. So the smallest value of n is 27.

(2) The required probability can be written in the form

$$P\left[\frac{n-1}{2} \leqslant V \leqslant \frac{3(n-1)}{2}\right] = P\left[V \leqslant \frac{3(n-1)}{2}\right] - P\left(V \leqslant \frac{n-1}{2}\right),$$

where $V=\frac{(n-1)S^2}{\sigma^2}$ has the χ^2-distribution with $n-1$ degrees of freedom. By trial and error, using computer program, it is found that for $n=12$, V has 11 degrees of freedom and $P(V \leqslant 16.5) - P(V \leqslant 5.5) = 0.876\,4 - 0.095\,4 = 0.781 < 0.8$. However, for $n=13$, V has 12 degrees of freedom and

$$P(V \leqslant 18) - P(V \leqslant 6) = 0.884\,3 - 0.083\,9 = 0.800\,4 > 0.8.$$

So the smallest value of n is 13.

Problem 6.29 Suppose population $X \sim N(\mu, \sigma^2)$, $(X_1, X_2, \cdots, X_n)$ is a sample from population X, $\overline{X}$ is the sample mean and S^2 is the sample variance. Determine

(1) the value of k such that $P(\mu - kS < \overline{X} < \mu + kS) = 0.9$;

(2) the value of a such that $P(S^2 > a) = 0.1$ when $\sigma = 5$.

Solution:

(1) $\dfrac{\overline{X}-\mu}{S/\sqrt{n}} \sim t(n-1)$,

$$0.9=P(\mu-kS<\overline{X}<\mu+kS)=P\left(-k\sqrt{n}\leqslant\frac{\overline{X}-\mu}{S/\sqrt{n}}\leqslant k\sqrt{n}\right),$$

$$P\left(\frac{\overline{x}-\mu}{s/\sqrt{n}}>k\sqrt{n}\right)=P[t(n-1)>k\sqrt{n}]=0.05, k\sqrt{n}=t_{0.05}(n-1),$$

therefore, $k=\dfrac{t_{0.05}(n-1)}{\sqrt{n}}$.

(2) $\dfrac{(n-1)S^2}{\sigma^2}\sim\chi^2(n-1)$, $0.1=P(S^2>a)=P\left[\dfrac{(n-1)S^2}{\sigma^2}>\dfrac{(n-1)a}{\sigma^2}\right]$,

$$\frac{(n-1)a}{\sigma^2}=\chi^2_{0.1}(n-1),$$

therefore,

$$a=\frac{\sigma^2\chi^2_{0.1}(n-1)}{n-1}=\frac{25\chi^2_{0.1}(n-1)}{n-1}.$$

Problem 6.30 Suppose that a random sample of 16 observations is drawn from the normal distribution with mean μ and standard deviation 12, and that independently another random sample of 25 observations is drawn from the normal distribution with the same mean μ and standard deviation 20. Let $\overline{X}$ and $\overline{Y}$ denote the sample means of the two samples. Evaluate $P(|\overline{X}-\overline{Y}|<5)$.

Solution:

$\overline{X}$ and $\overline{Y}$ have independent normal distributions with the same mean μ, and $V(\overline{X})=\dfrac{144}{16}=9$, $V(\overline{Y})=\dfrac{400}{25}=16$. Hence, $\overline{X}-\overline{Y}$ has the normal distribution with mean 0 and variance $9+16=25$. Thus, $Z=(\overline{X}-\overline{Y})/5$ has the standard normal distribution $N(0,1)$. It follows that the required probability is

$$P(|\overline{X}-\overline{Y}|<5)=P(|Z|<1)=2\Phi(1)-1=0.682\,6.$$

Problem 6.31 Suppose that a random sample is to be taken from the normal distribution with unknown mean μ and standard deviation 2. Let $\overline{X}$ be its sample mean.

(1) How large a random sample must be taken in order that $E(|\bar{X}-\mu|^2) \leqslant 0.1$ for every possible value of μ?

(2) How large a random sample must be taken in order that $E(|\bar{X}-\mu|) \leqslant 0.1$ for every possible value of μ?

 **Solution**:

(1) $\bar{X} \sim N\left(\mu, \frac{4}{n}\right), \bar{X}-\mu \sim N\left(0, \frac{4}{n}\right)$,

$0.1 \geqslant E(|\bar{X}-\mu|^2) = E[(\bar{X}-\mu)^2] = V(\bar{X}-\mu) + [E(\bar{X}-\mu)]^2 = \frac{4}{n}.$

Therefore, $n \geqslant 40$.

Another Method:

$\frac{\bar{X}-\mu}{2}\sqrt{n} \sim N(0,1), \frac{(\bar{X}-\mu)^2}{2^2}(\sqrt{n})^2 \sim \chi^2(1)$,

$$0.1 \geqslant E(|\bar{X}-\mu|^2) = \frac{4}{n}E\left[\frac{(\bar{X}-\mu)^2}{2^2}(\sqrt{n})^2\right] = \frac{4}{n}E(\chi^2(1)) = \frac{4}{n}.$$

Solve this inequality, we have $n \geqslant 40$.

(2) $\bar{X} \sim N\left(\mu, \frac{4}{n}\right), \bar{X}-\mu \sim N\left(0, \frac{4}{n}\right), Z = \frac{\bar{X}-\mu}{2/\sqrt{n}} \sim N(0,1)$,

$E(|\bar{X}-\mu|) = \frac{2}{\sqrt{n}}E(|Z|) = \frac{2}{\sqrt{n}}\int_{-\infty}^{+\infty}|z|\frac{1}{\sqrt{2\pi}}e^{-\frac{z^2}{2}}dz = \frac{2\sqrt{2}}{\sqrt{n\pi}}\int_{0}^{+\infty} z e^{-\frac{z^2}{2}}dz = \frac{2\sqrt{2}}{\sqrt{n\pi}}.$

Solve the inequality $\frac{2\sqrt{2}}{\sqrt{n\pi}} \leqslant 0.1$, we have $n \geqslant \frac{800}{\pi}, n \geqslant 255$.

Problem 6.32 Here are the lengths (unit: cm) for 16 different nails of a certain brand:

2.14, 2.10, 2.13, 2.15, 2.13, 2.12, 2.13, 2.10,

2.15, 2.12, 2.14, 2.10, 2.13, 2.11, 2.14, 2.11.

Assume that these numbers are the observed values from a random sample of 16 independent normal random variables with mean μ and variance σ^2. Find a 90% confidence interval for the mean length μ.

(1) $\sigma = 0.01$ (cm) is known; (2) σ is unknown.

 **Solution**:

(1) $n = 16, \bar{x} = \frac{1}{16}(2.14 + \cdots + 2.11) = 2.125, \alpha = 0.10, z_{\alpha/2} = z_{0.05} = 1.645$, the condfidence

interval of μ is

$$\left(\bar{x}-\frac{\sigma z_{\alpha/2}}{\sqrt{n}},\bar{x}+\frac{\sigma z_{\alpha/2}}{\sqrt{n}}\right)=\left(2.125-\frac{0.01\times1.645}{\sqrt{16}},2.125+\frac{0.01\times1.645}{\sqrt{16}}\right)=(2.121,2.129).$$

(2) $n=16$, $\bar{x}=\frac{1}{16}\times(2.14+\cdots+2.11)=2.125$, $s=0.017\ 13$, $\alpha=0.10$, $t_{\alpha/2}(n-1)=t_{0.05}(15)=1.753\ 1$, the condfidence interval of μ is $\left(\bar{x}-\frac{s\,t_{\alpha/2}(n-1)}{\sqrt{n}},\bar{x}+\frac{s\,t_{\alpha/2}(n-1)}{\sqrt{n}}\right)=\left(2.125-\frac{0.017\ 13\times1.753\ 1}{\sqrt{16}},2.125+\frac{0.017\ 13\times1.753\ 1}{\sqrt{16}}\right)=(2.117\ 5,2.132\ 5)$.

Problem 6.33 Suppose that $X_1,\cdots,X_n$ form a random sample from the normal distribution with unknown mean μ and known variance σ^2. How large a random sample must be taken in order that there will be a confidence interval for μ with confidence level 0.95 and its length is not larger than L?

Solution:

The variance σ^2 is known, then the confidence interval for μ is

$$\left(\bar{x}-\frac{\sigma z_{0.025}}{\sqrt{n}},\bar{x}+\frac{\sigma z_{0.025}}{\sqrt{n}}\right).$$

The length of the interval is $\frac{2\sigma z_{0.025}}{\sqrt{n}}$. Therefore, $\frac{2\sigma z_{0.025}}{\sqrt{n}}\leqslant L$, that is, $n\geqslant\frac{4\sigma^2 z_{0.025}^2}{L^2}$.

Problem 6.34 A random sample of 10 light bulbs has a mean lifetime of 1 500 hours and the sample standard deviation of the lifetime is 20 hours. Assume that the lifetime follows the normal distribution.

(1) Construct a 95% confidence interval estimate of the mean lifetime.

(2) Construct a 95% confidence interval estimate of the population standard deviation.

Solution:

(1) $n=10$, $\bar{x}=1\ 500$, $s=20$, $\alpha=0.05$, $t_{\alpha/2}(n-1)=t_{0.025}(9)=2.262\ 2$, the confidence interval for μ is

$$\left(\bar{x}-\frac{s\,t_{\alpha/2}(n-1)}{\sqrt{n}},\bar{x}+\frac{s\,t_{\alpha/2}(n-1)}{\sqrt{n}}\right)=\left(1\ 500-\frac{20\times2.262\ 2}{\sqrt{10}},1\ 500+\frac{20\times2.262\ 2}{\sqrt{10}}\right)=(1\ 485.693,1\ 514.307).$$

(2) $\chi^2_{\alpha/2}(n-1)=\chi^2_{0.025}(9)=19.022$, $\chi^2_{1-\alpha/2}(n-1)=\chi^2_{0.975}(9)=2.7$, the confidence interval for σ^2 is $\left(\frac{(n-1)s^2}{\chi^2_{\alpha/2}(n-1)},\frac{(n-1)s^2}{\chi^2_{1-\alpha/2}(n-1)}\right)=\left(\frac{9\times20^2}{19.022},\frac{9\times20^2}{2.7}\right)=(189.254\ 5,1\ 333.333)$.

The confidence interval for σ is $(\sqrt{189.2545}, \sqrt{1333.333}) = (13.757, 36.515)$.

Problem 6.35 Suppose the amount of drying time(unit: hour) of a particular paint follows $N(\mu, \sigma^2)$. The following data are 9 observations of a sample:

6.0, 5.7, 5.8, 6.5, 7.0, 6.3, 5.6, 6.1, 5.0

(1) Construct a 95% confidence interval estimate of the mean drying time.

(2) Construct a 95% confidence interval estimate of the population standard deviation.

Solution:

(1) $n=9, \bar{x}=\frac{1}{9}(6.0+\cdots+5.0)=6, s=0.5745, \alpha=0.05, t_{\alpha/2}(n-1)=t_{0.025}(8)=2.3060$, the confidence interval for μ is

$$\left(\bar{x}-\frac{s\,t_{\alpha/2}(n-1)}{\sqrt{n}}, \bar{x}+\frac{s\,t_{\alpha/2}(n-1)}{\sqrt{n}}\right)=\left(6-\frac{0.5745\times2.3060}{\sqrt{9}}, 6+\frac{0.5745\times2.3060}{\sqrt{9}}\right)=(5.558, 6.442).$$

(2) $\chi^2_{\alpha/2}(n-1)=\chi^2_{0.025}(8)=17.534, \chi^2_{1-\alpha/2}(n-1)=\chi^2_{0.975}(8)=2.18$, the confidence interval for σ^2 is

$$\left(\frac{(n-1)s^2}{\chi^2_{\alpha/2}(n-1)}, \frac{(n-1)s^2}{\chi^2_{1-\alpha/2}(n-1)}\right)=\left(\frac{8\times0.5745^2}{17.534}, \frac{8\times0.5745^2}{2.18}\right)=(0.1506, 1.2112).$$

The confidence interval for σ is $(\sqrt{0.1506}, \sqrt{1.2112}) = (0.388, 1.1)$.

Problem 6.36 A research team is interested in the difference between serum uric acid levels in patients with and without Down's syndrome. In a large hospital for the treatment of the mentally retarded, a sample of 12 individuals with Down's syndrome yielded a mean of $\bar{x}_1 = 4.5$ mg/100 mL and sample variance $s_1^2 = 4.9$. In a general hospital a sample of 15 normal individuals of the same age and sex were found to have a mean value of $\bar{x}_2 = 3.4$ mg/100 mL and sample variance $s_2^2 = 5.6$. If it is reasonable to assume that the two populations of values are normally distributed with equal variances, find the 95 percent confidence interval for $\mu_1-\mu_2$.

Solution:

$m=12, n=15, m+n-2=25, \bar{x}_1=4.5, s_1^2=4.9, \bar{x}_2=3.4, s_2^2=5.6, \alpha=0.05, t_{\frac{\alpha}{2}}(m+n-2)=t_{0.025}(25)=2.0595$.

$$s_p=\sqrt{\frac{(m-1)s_1^2+(n-1)s_2^2}{m+n-2}}=\sqrt{\frac{11\times4.9+14\times5.6}{25}}=2.30.$$

The 95 percent confidence interval for $\mu_1-\mu_2$ is

$$\left(\bar{x}_1-\bar{x}_2-t_{\frac{\alpha}{2}}(m+n-2)s_P\sqrt{\frac{1}{m}+\frac{1}{n}},\bar{x}_1-\bar{x}_2+t_{\frac{\alpha}{2}}(m+n-2)s_P\sqrt{\frac{1}{m}+\frac{1}{n}}\right)$$

$$=\left(4.5-3.4-2.059\ 5\times2.30\times\sqrt{\frac{1}{12}+\frac{1}{15}},4.5-3.4+2.059\ 5\times2.30\times\sqrt{\frac{1}{12}+\frac{1}{15}}\right)$$

$$\approx(-0.735,2.935).$$

Problem 6.37 Among 11 patients in a certain study, the standard deviation of the property of interest was 5.8. In another group of 4 patients, the standard deviation was 3.4.

(1) Construct a 95% confidence interval for the ratio of the variances of these two populations.

(2) Find the lower limit of a one-sided 95% confidence interval.

 **Solution**:

Let A denote the first group and let B be the second group. $m=11,n=4,s_A^2=5.8^2,s_B^2=3.4^2$.

(1) $\alpha=0.05$, using the F-table, it is found that

$$F_{\frac{\alpha}{2}}(m-1,n-1)=F_{0.025}(10,3)=14.419.$$

$$F_{1-\frac{\alpha}{2}}(m-1,n-1)=F_{0.975}(10,3)=\frac{1}{F_{0.025}(3,10)}=\frac{1}{4.826}=0.207.$$

A 95% confidence interval for σ_A^2/σ_B^2 is

$$\left(\frac{s_A^2}{s_B^2F_{\frac{\alpha}{2}}(m-1,n-1)},\frac{s_A^2}{s_B^2F_{1-\frac{\alpha}{2}}(m-1,n-1)}\right)\approx(0.202,14.058).$$

(2) $\alpha=0.05,F_\alpha(m-1,n-1)=F_{0.05}(10,3)=8.786.$

$$F=\frac{S_A^2}{\sigma_A^2}\Big/\frac{S_B^2}{\sigma_B^2}\sim F(m-1,n-1).$$

For confidence level $1-\alpha$,

$$1-\alpha=P\left(\frac{S_A^2}{\sigma_A^2}\Big/\frac{S_B^2}{\sigma_B^2}<F_\alpha(m-1,n-1)\right)$$

$$=P\left(\frac{\sigma_A^2}{\sigma_B^2}>\frac{S_A^2}{S_B^2F_\alpha(m-1,n-1)}\right).$$

The 95% one-sided confidence lower limit is

$$\frac{s_A^2}{s_B^2F_\alpha(m-1,n-1)}\approx0.331\ 2.$$

Problem 6.38 Find a 99% confidence interval for the proportion of heads obtained if 175 heads were counted after 400 tosses of a coin. Is this coin balanced?

Solution:

Let $X_i=\begin{cases}1 & \text{the } i\text{-th toss is head}\\ 0 & \text{the } i\text{-th toss is tail}\end{cases}$, then $X_1, \cdots, X_{400}$ form a sample from the Bernoulli distribution $X \sim B(1,p)$.

$n=400, \bar{x}=\dfrac{175}{400}=0.437\,5, \alpha=0.01$, it is found from the standard normal table that $z_{\alpha/2}=z_{0.005}=2.575, z_{\alpha/2}^2=z_{0.005}^2=6.631$.

$a=z_{\alpha/2}^2+n=406.631, b=-(2n\bar{x}+z_{\alpha/2}^2)=-356.631, c=n\bar{x}^2=76.562\,5$.

Therefore,

$$p_1=\frac{1}{2a}(-b-\sqrt{b^2-4ac})=0.375\,2, p_2=\frac{1}{2a}(-b+\sqrt{b^2-4ac})=0.501\,9.$$

A 99% confidence interval for the proportion of heads is $(0.375\,2, 0.501\,9)$. The interval contains 0.5, and the coin is balanced.

Problem 6.39 Suppose $X \sim P(\lambda)$, where $\lambda>0$ is a parameter. $(X_1, \cdots, X_n)$ is a sample from the population X. Construct a $1-\alpha(0<\alpha<1)$ confidence interval for λ.

Solution:

By the law of large numbers, as n goes to infinity, the random variable $\dfrac{\sum\limits_{i=1}^{n} X_i - n\lambda}{\sqrt{n\lambda}}$ approximately follows $N(0,1)$, that is,

$$Z=\frac{\bar{X}-\lambda}{\sqrt{\dfrac{\lambda}{n}}} \sim N(0,1).$$

For $0<\alpha<1$, $z_{\alpha/2}$ can be found in the standard normal table, and satisfies

$$P\left(\left|\frac{\bar{X}-\lambda}{\sqrt{\dfrac{\lambda}{n}}}\right|<z_{\alpha/2}\right)=1-\alpha.$$

The inequality $\left|\dfrac{\bar{X}-\lambda}{\sqrt{\dfrac{\lambda}{n}}}\right|<z_{\alpha/2}$ is equivalent to $\lambda^2-\left(2\bar{X}+\dfrac{z_{\alpha/2}^2}{n}\right)\lambda+\bar{X}^2<0$.

Let

$$\lambda_1=\frac{2\bar{X}+\dfrac{z_{\alpha/2}^2}{n}-\sqrt{\left[2\bar{X}+\dfrac{z_{\alpha/2}^2}{n}\right]^2-4\bar{X}^2}}{2},$$

$$\lambda_2=\frac{2\overline{X}+\frac{z_{\alpha/2}^2}{n}+\sqrt{\left[2\overline{X}+\frac{z_{\alpha/2}^2}{n}\right]^2-4\overline{X}^2}}{2}.$$

A $1-\alpha(0<\alpha<1)$ confidence interval for λ is (λ_1,λ_2).

Problem 6.40 Suppose the population follows the exponential distribution with its PDF

$$f(x)=\begin{cases}\lambda e^{-\lambda x}, & x\geqslant 0\\ 0, & x<0\end{cases},$$

where $\lambda>0$ is a parameter. $(X_1,\cdots,X_n)$ is a sample from the population X. Find the lower limit of a one-sided $1-\alpha(0<\alpha<1)$ confidence interval for λ.

Solution:

$X_i\sim E(\lambda)\ (i=1,2,\cdots,n)$ are IID, then $X_1+\cdots+X_n\sim \text{Gamma}\left(\alpha=n,\beta=\frac{1}{\lambda}\right)$ with the following PDF

$$f(x)=\begin{cases}\frac{1}{\beta^{\alpha}\Gamma(\alpha)}x^{\alpha-1}e^{-\frac{x}{\beta}}, & x>0\\ 0, & x\leqslant 0\end{cases}=\begin{cases}\frac{\lambda^n}{(n-1)!}x^{n-1}e^{-\lambda x}, & x>0\\ 0, & x\leqslant 0\end{cases}.$$

Let $Y=2n\lambda\overline{X}=2\lambda(X_1+\cdots+X_n)$, then its PDF is

$$f(y)=\begin{cases}\frac{\lambda^n}{(n-1)!}\left(\frac{y}{2\lambda}\right)^{n-1}e^{-\lambda\left(\frac{y}{2\lambda}\right)}\times\left|\frac{1}{2\lambda}\right|, & y>0\\ 0, & y\leqslant 0\end{cases}=\begin{cases}\frac{y^{n-1}}{2^n(n-1)!}e^{-\frac{y}{2}}, & y>0\\ 0, & y\leqslant 0\end{cases}.$$

It is the PDF of $\chi^2(2n)$. That is, $2n\lambda\overline{X}\sim\chi^2(2n)$.

Another method to find the distribution of $2n\lambda\overline{X}$ is as follows: Consider the PDF of $Y=2\lambda X$

$$f(y)=\begin{cases}\lambda e^{-\lambda\left(\frac{y}{2\lambda}\right)}\times\left|\frac{1}{2\lambda}\right|, & y>0\\ 0, & y\leqslant 0\end{cases}=\begin{cases}\frac{1}{2}e^{-\frac{y}{2}}, & y>0\\ 0, & y\leqslant 0\end{cases}\sim\chi^2(2).$$

By the additive property of the Chi-square distribution,

$$2n\lambda\overline{X}=2\lambda X_1+\cdots+2\lambda X_n\sim\chi^2(2n).$$

Therefore,

$$1-\alpha=P[2n\lambda\overline{X}>\chi^2_{1-\alpha}(2n)]=P\left[\lambda>\frac{\chi^2_{1-\alpha}(2n)}{2n\overline{X}}\right].$$

The lower limit of a one-sided $1-\alpha(0<\alpha<1)$ confidence interval for λ is $\frac{\chi^2_{1-\alpha}(2n)}{2n\overline{X}}$.

Chapter 7 Hypothesis Testing

Summary of Knowledge

Exercise Solutions

Problem 7.1 Suppose $x_1, x_2, \cdots, x_n$ form a sample from $N(\mu, 1)$. Consider the following hypothesis testing problem:

$$H_0: \mu=2, H_1: \mu=3.$$

The rejection region $W=\{\bar{x} \geqslant 2.6\}$.

(1) If $n=20$, find the probabilities of two types of errors;

(2) What is the smallest sample size of n such that the probability of type II error satisfies $\beta \leqslant 0.01$?

(3) Show that if $n \to +\infty$, then $\alpha \to 0, \beta \to 0$.

Solution:

(1) The probability of type I error is

$$\begin{aligned} P(\text{Reject } H_0 \mid H_0 \text{ is true}) &= P(\bar{X} \geqslant 2.6 \mid \mu=2) \\ &= P\left(\frac{\bar{X}-2}{1/\sqrt{20}} \geqslant \frac{2.6-2}{1/\sqrt{20}}\right) \\ &= 1-\Phi(2.68) \\ &= 0.003\ 6. \end{aligned}$$

The probability of type II error is

$$\begin{aligned} P(\text{Accept } H_0 \mid H_1 \text{is true}) &= P(\bar{X}<2.6 \mid \mu=3) \\ &= P\left(\frac{\bar{X}-3}{1/\sqrt{20}} < \frac{2.6-3}{1/\sqrt{20}}\right) \\ &= \Phi(-1.79) \\ &= 0.036\ 7. \end{aligned}$$

(2) $\Phi\left(\frac{2.6-3}{1/\sqrt{n}}\right)\leqslant 0.01$, $\frac{2.6-3}{1/\sqrt{n}}\leqslant -2.33$, $n\geqslant 33.9$, the smallest value of n is 34.

(3) If $n\to+\infty$, the probability of type I error $\alpha=1-\Phi\left(\frac{2.6-2}{1/\sqrt{n}}\right)=1-\Phi(0.6\sqrt{n})\to 0$, the probability of type II error $\beta=\Phi\left(\frac{2.6-3}{1/\sqrt{n}}\right)=\Phi(-0.4\sqrt{n})\to 0$.

Problem 7.2 To test whether a coin is fair or not, flip a coin 10 times. Let X be the number of times that the coin comes up heads. If $|X-5|>3$, then reject $H_0: p=0.5$. What is the significance level of the test?

Solution:

The significance level of the test is

$$\begin{aligned}
&P(\text{Reject } H_0 \mid H_0 \text{ is true})\\
&=P(|X-5|>3 \mid p=0.5)\\
&=P(X<2 \text{ or } X>8 \mid p=0.5)\\
&=C_{10}^{0}\times0.5^{0}\times0.5^{10}+C_{10}^{1}\times0.5^{1}\times0.5^{9}+C_{10}^{9}\times0.5^{9}\times0.5^{1}+C_{10}^{10}\times0.5^{10}\times0.5^{0}\\
&=0.021\ 5.
\end{aligned}$$

Problem 7.3 Suppose $x_1, x_2, \cdots, x_n$ form a sample from the uniform distribution $U(0,\theta)$. Consider the following hypothesis testing problem:

$$H_0: \theta\geqslant 3, H_1: \theta<3$$

The rejection region $W=\{x_{(n)}\leqslant 2.5\}$.

(1) Find the largest value of α, the probability of type I error.

(2) What is the smallest value of sample size n such that the value in (1) is no more than 0.05?

Solution:

(1) The PDF of $X\sim U(0,\theta)$ is

$$f(x)=\begin{cases}\frac{1}{\theta}, & 0<x<\theta\\ 0, & \text{otherwise}\end{cases}.$$

The CDF of $X_{(n)}$ is

$$F_{X_{(n)}}(x)=[F(x)]^n=\begin{cases}0, & x\leqslant 0\\ \left(\frac{x}{\theta}\right)^n, & 0<x<\theta\\ 1, & x\geqslant\theta\end{cases}.$$

The probability of type I error is

$$P(\text{Reject } H_0 \mid H_0 \text{ is true}) = P(X_{(n)} \leqslant 2.5 \mid \theta \geqslant 3)$$
$$= F_{X_{(n)}}(2.5)$$
$$= \left(\frac{2.5}{\theta}\right)^n$$
$$\leqslant \left(\frac{2.5}{3}\right)^n.$$

(2) If the probability of type I error is no more than 0.05, then

$$\left(\frac{2.5}{3}\right)^n \leqslant 0.05.$$

Thus, $n \geqslant \dfrac{\ln 0.05}{\ln(2.5/3)} = 16.4$, the smallest value of n is 17.

Problem 7.4 Suppose $x_1, x_2, \cdots, x_n$ form a sample from population $X \sim N(\mu, \sigma^2)$, where σ^2 is known, μ takes one of the two values, μ_0 or $\mu_1(\mu_1 > \mu_0)$. $\bar{x}$ is the sample mean. Under the significance level α, let's consider

$$H_0: \mu = \mu_0, H_1: \mu = \mu_1.$$

The probability of the type II error is

$$\beta = P\{\bar{X} - \mu_0 < k \mid \mu = \mu_1\}.$$

Show that

$$\beta = \Phi\left(z_\alpha - \frac{\mu_1 - \mu_0}{\sigma/\sqrt{n}}\right)$$

and

$$n = (z_\alpha + z_\beta)^2 \frac{\sigma^2}{(\mu_1 - \mu_0)^2}.$$

 **Solution:**

The probability of type I error is

$$\alpha = P(\text{Reject } H_0 \mid H_0 \text{ is true})$$
$$= P\{\bar{X} - \mu_0 \geqslant k \mid \mu = \mu_0\}$$
$$= P\left(\frac{\bar{X} - \mu_0}{\sigma/\sqrt{n}} \geqslant \frac{k}{\sigma/\sqrt{n}}\right)$$
$$= 1 - \Phi\left(\frac{k}{\sigma/\sqrt{n}}\right).$$

Thus, $z_\alpha = \dfrac{k}{\sigma/\sqrt{n}}, k = \dfrac{\sigma z_\alpha}{\sqrt{n}}$.

The probability of type II error β is

$$\begin{aligned}\beta &= P(\text{Accept } H_0 \mid H_1 \text{is true}) \\ &= P\{\overline{X}-\mu_0<k \mid \mu=\mu_1\} \\ &= P\left(\frac{\overline{X}-\mu_1}{\sigma/\sqrt{n}}<\frac{k+\mu_0-\mu_1}{\sigma/\sqrt{n}}\right) \\ &= \Phi\left(z_\alpha-\frac{\mu_1-\mu_0}{\sigma/\sqrt{n}}\right).\end{aligned}$$

Thus, $z_\beta=\frac{\mu_1-\mu_0}{\sigma/\sqrt{n}}-z_\alpha$, that is,

$$z_\alpha+z_\beta=\frac{\mu_1-\mu_0}{\sigma/\sqrt{n}}, n=(z_\alpha+z_\beta)^2\frac{\sigma^2}{(\mu_1-\mu_0)^2}.$$

Note: (1) if n is fixed, as α decreases, then z_α increases, z_β decreases, β increases.

(2) if n is fixed, as β decreases, then z_β increases, z_α decreases, α increases.

Problem 7.5 A simple random sample of 10 people from a certain population has a mean age of 27 years old. The variance is known to be 20 years^2. Let $\alpha=0.05$.

(1) Can we conclude that the mean age of the population is not 30 years old?

(2) Can we conclude that the mean age of the population is less than 30 years old?

Solution:

(1) Let μ be the mean age of the population. The hypotheses to be tested under the significance level 0.05 are

$$H_0: \mu=30, H_1: \mu\neq 30.$$

Since the variance of the population $\sigma^2=20$ is known, we use the test statistic

$$Z=\frac{\overline{X}-\mu_0}{\frac{\sigma}{\sqrt{n}}}.$$

The rejection region is $|z|=\left|\frac{\overline{x}-\mu_0}{\frac{\sigma}{\sqrt{n}}}\right|>z_{0.025}=1.96$.

$n=10, \overline{x}=27, \sigma^2=20$, the observation of $|z|$ is $|z|=\left|\frac{27-30}{\sqrt{\frac{20}{10}}}\right|=2.12>1.96$.

The sample data fall in the rejection region and H_0 should be rejected. We can conclude that the mean age of the population is not 30 years old.

(2) The hypotheses to be tested at the significance level 0.05 are

$$H_0: \mu=30, H_1: \mu<30.$$

Since the variance of the population $\sigma^2=20$ is known, we use the test statistic

$$Z=\frac{\overline{X}-\mu_0}{\frac{\sigma}{\sqrt{n}}}.$$

The rejection region is $z=\frac{\overline{x}-\mu_0}{\frac{\sigma}{\sqrt{n}}}<-z_{0.05}=-1.645$.

$n=10, \overline{x}=27, \sigma^2=20$, the observation of z is $z=\frac{27-30}{\sqrt{20}/\sqrt{10}}=-2.12<-1.645$.

The sample data fall in the rejection region and H_0 should be rejected. We can conclude that the mean age of the population is less than 30 years old.

Problem 7.6 A rental car company claims the mean time to rent a car on their website is 60 seconds with a standard deviation of 30 seconds. A random sample of 36 customers attempted to rent a car on the website. The mean time to rent was 75 seconds. Is this enough evidence to contradict the company's claim? ($\alpha=0.05$)

 **Solution 1:**

Let μ be the mean time to rent a car. The hypotheses to be tested at the significance level 0.05 are

$$H_0: \mu\leqslant 60, H_1: \mu>60.$$

Since the variance of the population $\sigma^2=30^2$ is known, we use the test statistic

$$Z=\frac{\overline{X}-\mu_0}{\frac{\sigma}{\sqrt{n}}}.$$

The rejection region is $z=\frac{\overline{x}-\mu_0}{\frac{\sigma}{\sqrt{n}}}>z_{0.05}=1.645$.

$n=36, \overline{x}=75, \sigma=30$, the observation of z is $z=\frac{75-60}{30/\sqrt{36}}=3>1.645$.

The sample data fall in the rejection region and H_0 should be rejected. We can conclude that the mean time to rent a car is larger than 60 minutes. We conclude that this is enough evidence to question their claim.

 **Solution 2:**

We can also adopt the p-value approach to the hypothesis testing. The z-score is

$$z=\frac{75-60}{30/\sqrt{36}}=3$$

The p-value is the probability that we obtain the observed value of the test statistic that is more extreme in the direction of the alternative hypothesis, calculated when H_0 is true, as shown in Figure 7.1.

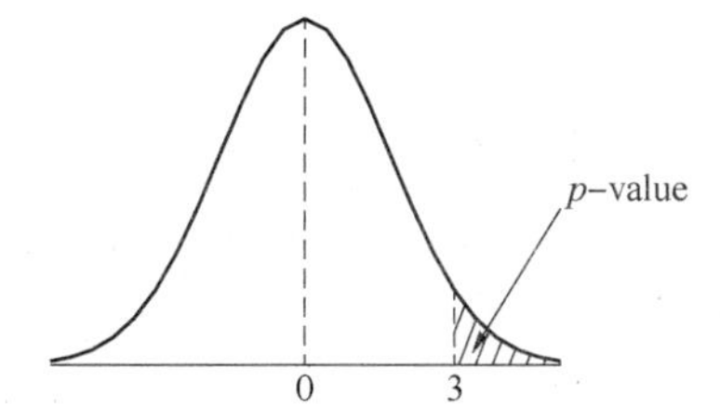

Figure 7.1

$$p=P(Z>3)=1-\Phi(3)=0.001\ 3.$$

$\alpha=0.05$, p-value $=0.001\ 3$, p-value $<\alpha$, we should reject H_0. We can conclude that the mean time to rent a car is larger than 60 minutes. We conclude that this is enough evidence to question their claim. The conclusion is identical to that obtained by the critical value method.

Problem 7.7 A certain type of electronic product is considered to be acceptable if its lifetime is at least 1 000 hours. Draw a sample of size 25 of these products and the sample mean is 950 hours. Assume the lifetime follows the normal distribution $N(\mu,\sigma^2)$ and $\sigma=100$. Are the products acceptable? Use $\alpha=0.05$.

 **Solution 1:**

Let μ be the mean lifetime of the electronic product. The hypotheses to be tested at the significance level 0.05 are

$$H_0: \mu \geqslant 1\ 000, H_1: \mu<1\ 000.$$

Since the variance of the population $\sigma^2=100^2$ is known, we use the test statistic

$$Z=\frac{\overline{X}-\mu_0}{\frac{\sigma}{\sqrt{n}}}.$$

The rejection region is $z=\frac{\overline{x}-\mu_0}{\frac{\sigma}{\sqrt{n}}}<-z_{0.05}=-1.645$.

$n=25$, $\overline{x}=950$, $\sigma=100$, the observation of z is $z=\frac{950-1\ 000}{100/\sqrt{25}}=-2.5<-1.645$.

The sample data fall in the rejection region and H_0 should be rejected. We conclude that the products are not acceptable.

Solution 2:

We can also adopt the p-value approach to the hypothesis testing. The z-score is

$$z=\frac{950-1\ 000}{100/\sqrt{25}}=-2.5.$$

The p-value is the probability that we obtain the observed value of the test statistic that is more extreme in the direction of the alternative hypothesis, calculated when H_0 is true, as shown in Figure 7.2.

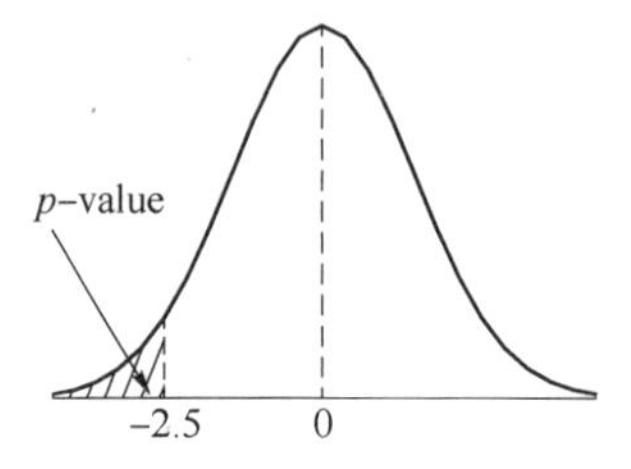

Figure 7.2

$$p=P(Z<-2.5)=1-\Phi(2.5)=0.006\ 2.$$

$\alpha=0.05$, p-value $=0.006\ 2$, p-value $<\alpha$, we should reject H_0. We conclude that the products are not acceptable. The conclusion is identical to that obtained by the critical value method.

Problem 7.8 Signals are sent from location A to location B. The signal received at location B follows $N(\mu,0.2^2)$, where μ is the true signal value sent at location A. Suppose the signal is sent five times, and the received values are 8.05, 8.15, 8.2, 8.1, 8.25, can we conclude that the true signal value sent is 8? Use $\alpha=0.05$.

Solution 1:

The hypotheses to be tested at the significance level 0.05 are

$$H_0:\mu=8, H_1:\mu\neq 8.$$

Since the variance of the population $\sigma^2=0.2^2$ is known, we use the test statistic

$$Z=\frac{\overline{X}-\mu_0}{\frac{\sigma}{\sqrt{n}}}.$$

The rejection region is $|z|=\left|\frac{\overline{x}-\mu_0}{\frac{\sigma}{\sqrt{n}}}\right|>z_{\frac{\alpha}{2}}=z_{0.025}=1.96.$

$n=5$, $\overline{x}=\frac{8.05+8.15+8.2+8.1+8.25}{5}=8.15$, $\sigma=0.2$, and $|z|=\frac{8.15-8}{0.2/\sqrt{5}}=1.677<1.96.$

The sample data fall outside the rejection region and we fail to reject H_0. The sample data do not rule out the possibility that the true signal value sent is 8.

Solution 2:

We can also adopt the p-value approach to the hypothesis testing. The z-score is

$$z=\frac{8.15-8}{0.2/\sqrt{5}}=1.68.$$

The p-value is the probability that we obtain the observed value of the test statistic that is more extreme in the direction of the alternative hypothesis, calculated when H_0 is true, as shown in Figure 7.3.

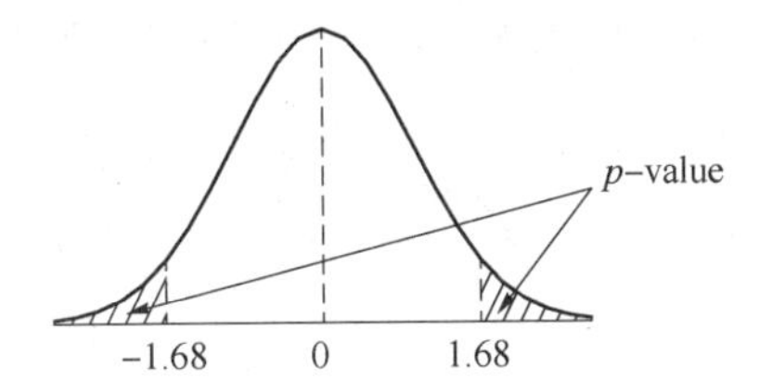

Figure 7.3

$$p=P(|Z|\geqslant 1.68)=2-2\Phi(1.68)=0.093.$$

$\alpha=0.05$, p-value $=0.093$, p-value $>\alpha$, we do not reject H_0. The sample data do not rule out the possibility that the true signal value sent is 8. The conclusion is identical to that obtained by the critical value method.

Problem 7.9 When the cutter works properly, the length of each piece of metal rod produced is a random variable that obeys the normal distribution $X\sim N(\mu,\sigma^2)$, $\mu=10.5$ cm, $\sigma=0.15$ cm. From a batch of products, randomly select 15 metal rods, their lengths (unit: cm) are as follows:

10.4, 10.6, 10.1, 10.4, 10.5, 10.3, 10.3, 10.2, 10.9, 10.6, 10.8, 10.5, 10.7, 10.2, 10.7.

Does the cutter work properly? Use $\alpha=0.05$.

Solution 1:

The hypotheses to be tested at the significance level 0.05 are

$$H_0: \mu=10.5, H_1: \mu\neq 10.5.$$

Since the variance of the population $\sigma^2=0.15^2$ is known, we use the test statistic

$$Z=\frac{\overline{X}-\mu_0}{\frac{\sigma}{\sqrt{n}}}.$$

The rejection region is $|z|=\left|\frac{\overline{x}-\mu_0}{\frac{\sigma}{\sqrt{n}}}\right|>z_{\frac{\alpha}{2}}=z_{0.025}=1.96.$

$$n=15,\bar{x}=\frac{10.4+10.6+\cdots+10.7}{15}=10.48,\sigma=0.15,\text{ and } |z|=\left|\frac{10.48-10.5}{0.15/\sqrt{15}}\right|=0.516<1.96.$$

The sample data fall outside the rejection region and we fail to reject H_0. The sample data do not rule out the possibility that the cutter works properly.

Solution 2:

We can also adopt the p-value approach to hypothesis testing. The z-score is

$$z=\frac{10.48-10.5}{0.15/\sqrt{15}}=-0.52.$$

The p-value is the probability that we obtain the observed value of the test statistic that is more extreme in the direction of the alternative hypothesis, calculated when H_0 is true, as shown in Figure 7.4.

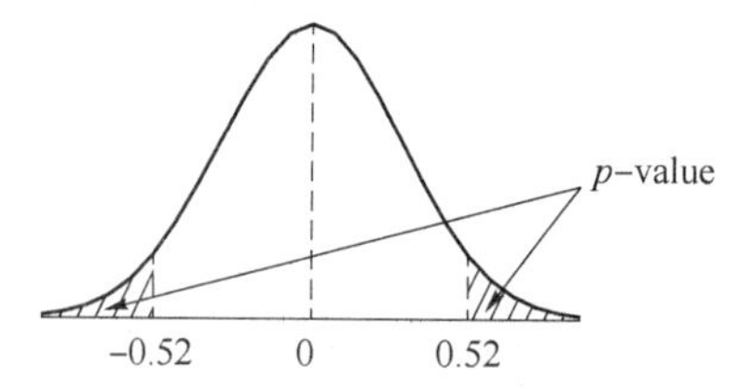

Figure 7.4

$$p=P(|Z|\geqslant|-0.52|)=2-2\Phi(0.52)=0.603.$$

$\alpha=0.05$, p-value $=0.603$, p-value $>\alpha$, we do not reject H_0. The sample data do not rule out the possibility that the cutter works properly. The conclusion is identical to that obtained by the critical value method.

Problem 7.10 It is regulated that the content of vitamin C in a certain kind of food per 100 g is not less than 21 mg. Assume the determination of vitamin C content obeys the normal distribution $N(\mu,\sigma^2)$. Randomly select a sample of size 17 from this batch of food, vitamin C content(unit: mg) is measured and the data are

16, 22, 21, 20, 23, 21, 19, 15, 13, 23, 17, 20, 29, 18, 22, 16, 25.

Try to test whether the content of vitamin C in this kind of food is qualified using $\alpha=0.05$.

Solution:

The hypotheses to be tested at the significance level 0.05 are

$$H_0:\mu\geqslant 21, H_1:\mu<21.$$

Since the variance of the population is unknown, we use the test statistic

$$T=\frac{\bar{X}-\mu_0}{S/\sqrt{n}}.$$

$n=17, \bar{x}=\dfrac{16+22+\cdots+25}{17}=20, s^2=15.875, s=3.984, \alpha=0.05, t_\alpha(n-1)=t_{0.05}(16)=1.746$. The rejection region is

$$t=\frac{\bar{x}-\mu_0}{s/\sqrt{n}}<-t_{0.05}(16)=-1.746.$$

The value of t is $t=\dfrac{20-21}{3.984/\sqrt{17}}=-1.035>-1.746$.

The sample data fall outside the rejection region and we fail to reject H_0. The sample data do not rule out the possibility that the content of vitamin C in this kind of food is qualified.

Problem 7.11 The strength of a steel of a certain type is known to obey the normal distribution. Based on past experience, the average tensile strength is 52.00(unit: kg/mm^2). The formulation of the steelmaking is improved, the new method is used to refine the steel, select 7 steel bars produced and the strengths are measured as follows:

$$52.45, 48.51, 56.02, 51.53, 49.02, 53.38, 54.04.$$

Is the mean strength of steel bar produced by the new method improved significantly? ($\alpha=0.05$)

Solution:

The hypotheses to be tested at the significance level 0.05 are

$$H_0: \mu=52, H_1: \mu>52.$$

Since the variance of the population is unknown, we use the test statistic

$$T=\frac{\bar{X}-\mu_0}{S/\sqrt{n}}.$$

$n=7, \bar{x}=\dfrac{52.45+48.51+\cdots+54.04}{7}=52.1357, s^2=7.2635, s=2.6951, \alpha=0.05, t_\alpha(n-1)=t_{0.05}(6)=1.943$. The rejection region is

$$t=\frac{\bar{x}-\mu_0}{s/\sqrt{n}}>t_{0.05}(6)=1.943.$$

The value of t is $\dfrac{52.1357-52}{2.6951/\sqrt{7}}=0.1332<1.943$. The sample data fall outside the rejection region and we fail to reject H_0. The sample data do not rule out the possibility that the mean strength of steel bar produced by the new method is not improved significantly.

Problem 7.12 A simple random sample of 14 people from a certain population gives body mass indices(BMIs). The data (unit: kg/m^2) are 23, 25, 21, 37, 39, 21, 23, 24, 32, 57, 23, 26, 31, 45. Can we conclude that the BMI is not 35 kg/m^2? ($\alpha=0.05$)

Solution:

The hypotheses to be tested at the significance level 0.05 are

$$H_0:\mu=35, H_1:\mu\neq 35.$$

Since the variance of the population is unknown, we use the test statistic

$$T=\frac{\overline{X}-\mu_0}{S/\sqrt{n}}.$$

$n=14, \bar{x}=\frac{23+25+\cdots+45}{14}=30.5, s^2=113.192, s=10.639, \alpha=0.05, t_{\frac{\alpha}{2}}(n-1)=t_{0.025}(13)=$ 2.16. The rejection region is

$$|t|=\left|\frac{\bar{x}-\mu_0}{s/\sqrt{n}}\right|>t_{0.025}(13)=2.16.$$

The value of of $|t|=\frac{30.5-35}{10.639/\sqrt{14}}=1.583<2.16.$

The sample data fall outside the rejection region and we fail to reject H_0. The sample data do not rule out the possibility that the BMI is 35 kg/m^2.

Problem 7.13 Researchers studying the effects of diet on growth would like to know if a vegetarian diet affects the height of a child. The researchers randomly selected 12 vegetarian children that are six years old. The average height of the children is 42.5 inches with a standard deviation of 3.8 inches. The average height for all six-year-old children is 45.75 inches. Conduct a hypothesis test to determine whether there is overwhelming evidence at $\alpha=0.05$ that six-year-old vegetarian children are not the same height as other six-year-old children. Assume that the heights of six-year-old vegetarian children are approximately normally distributed.

Solution:

The hypotheses to be tested at the significance level 0.05 are

$$H_0:\mu=45.75, H_1:\mu\neq 45.75.$$

Since the variance of the population is unknown, we use the test statistic

$$T=\frac{\overline{X}-\mu_0}{S/\sqrt{n}}.$$

$n=12, \bar{x}=42.5, s=3.8, \alpha=0.05, t_{\frac{\alpha}{2}}(n-1)=t_{0.025}(11)=2.201$. The rejection region is

$$|t|=\left|\frac{\bar{x}-\mu_0}{s/\sqrt{n}}\right|>t_{0.025}(11)=2.201.$$

The value of $|t|=\frac{42.5-45.75}{3.8/\sqrt{12}}=2.963>2.201.$

The sample data fall in the rejection region and we reject H_0. The sample data indicate that six-year-old vegetarian children are not the same height as other six-year-old children.

Problem 7.14 A travel writer suspects that hotels have higher room rates in Florida than in Georgia. He carried out a survey of room rates and the results are given below. Suppose the population standard deviations of room rates for Florida and Georgia are $\sigma_1=18$ and $\sigma_2=24$, respectively. (unit: dollar)

Group	Count	Mean
Florida	14	\$ 112
Georgia	26	\$ 109

Is there evidence that the mean room rate in Florida is higher than the mean room rate in Georgia? Conduct an appropriate hypothesis test using $\alpha=0.05$.

 **Solution**:

Let X denote the room rates in Florida, $X\sim N(\mu_1,18^2)$, and let Y denote the room rates in Georgia, $Y\sim N(\mu_2,24^2)$. The hypotheses to be tested at the significance level 0.05 are

$$H_0:\mu_1-\mu_2=0, H_1:\mu_1-\mu_2>0.$$

We use the test statistic

$$Z=\frac{\overline{X}-\overline{Y}}{\sqrt{\frac{\sigma_1^2}{n_1}+\frac{\sigma_2^2}{n_2}}}.$$

$n_1=14, n_2=26, \bar{x}=112, \bar{y}=109, \alpha=0.05, z_\alpha=z_{0.05}=1.645.$

The rejection region is

$$z>z_{0.05}=1.645.$$

The value of $z=\frac{112-109}{\sqrt{\frac{18^2}{14}+\frac{24^2}{26}}}=0.446<1.96$. The sample data fall outside the rejection region, so we fail to reject the null hypothesis. The sample data do not rule out the possibility that there are no significant differences in the mean room rate in Florida and Georgia.

Problem 7.15 The amount of a certain trace element in blood is known to vary with a standard deviation of 14.1 ppm (parts per million) for male blood donors and 9.5 ppm for female donors. Random samples of 75 male and 50 female donors yield concentration means of 28 and 33 ppm , respectively. Are the population means of concentrations of the element the same for men and women? ($\alpha=0.05$)

Solution:

Let X denote the amount of a certain trace element in blood for male blood donors, $X \sim N(\mu_1, 14.1^2)$, and let Y denote the the amount of a certain trace element in blood for female blood donors, $Y \sim N(\mu_2, 9.5^2)$.

The hypotheses to be tested at the significance level 0.05 are

$$H_0: \mu_1 - \mu_2 = 0, H_1: \mu_1 - \mu_2 \neq 0.$$

We use the test statistic

$$Z = \frac{\overline{X} - \overline{Y}}{\sqrt{\frac{\sigma_1^2}{n_1} + \frac{\sigma_2^2}{n_2}}}.$$

$n_1 = 75, n_2 = 50, \bar{x} = 28, \bar{y} = 33, \alpha = 0.05, z_{\frac{\alpha}{2}} = z_{0.025} = 1.96.$

The rejection region is

$$|z| > z_{0.025} = 1.96.$$

The value of $|z| = \left| \frac{28-33}{\sqrt{\frac{14.1^2}{75} + \frac{9.5^2}{50}}} \right| = 2.37 > 1.96$. The sample data fall in the rejection region, so we reject the null hypothesis H_0. The sample data indicate that the population means of concentrations of the element are not the same for men and women.

Problem 7.16 A grocery has two stores. One store is located on First Street and the other on Main Street and each is run by a different manager. Each manager claims that her store's layout maximizes the amounts customers will purchase on impulse. Both managers surveyed a sample of their customers and asked them how much more they spent than they had planned to, in other words, how much did they spend on impulse? The following table shows the sample data collected from the two stores. (unit: dollar)

First street: 15.78, 17.73, 10.61, 15.79, 14.22, 13.82, 13.45, 12.86, 10.82, 12.85

Main street: 15.19, 18.22, 15.38, 15.96, 21.92, 12.87, 12.47, 13.96, 13.79, 13.74, 18.4, 18.57, 17.79, 10.83.

Is there a difference in the mean amounts purchased on impulse at the two stores? Assume the populations to be approximately normal with equal variances. ($\alpha = 0.05$)

Solution:

Let X denote the amounts purchased on impulse at the First street store, $X \sim N(\mu_1, \sigma^2)$, and let Y denote the amounts purchased on impulse at the Main street store, $Y \sim N(\mu_2, \sigma^2)$.

The hypotheses to be tested at the significance level 0.05 are

$$H_0:\mu_1-\mu_2=0, H_1:\mu_1-\mu_2\neq 0.$$

The test statistic is

$$T=\frac{\overline{X}-\overline{Y}}{S_p\sqrt{\frac{1}{n_1}+\frac{1}{n_2}}}.$$

$n_1=10$, $n_2=14$, $\overline{x}=\frac{15.78+\cdots+12.85}{10}=13.793$, $\overline{y}=\frac{15.19+\cdots+10.83}{14}=15.649$, $s_p=\sqrt{\frac{(n_1-1)s_1^2+(n_2-1)s_2^2}{n_1+n_2-2}}=\sqrt{\frac{9\times 4.94+13\times 9.077}{22}}=2.717\ 46$, $\alpha=0.05$, $t_{\alpha/2}(n_1+n_2-2)=t_{0.025}(22)=2.073\ 9$.

The rejection region of the test is

$$|t|>t_{0.025}(22)=2.073\ 9.$$

The value of $|t|=\left|\frac{13.793-15.649}{2.717\ 46\times\sqrt{\frac{1}{10}+\frac{1}{14}}}\right|=1.649\ 6<2.073\ 9.$

The sample data fall outside the rejection region, so we fail to reject the null hypothesis. The sample data do not rule out the possibility that there are no significant differences in the mean amounts purchased on impulse at the two stores.

Problem 7.17 The following gives observations of the running time (unit: hour) after charging of two types of calculators:

Model A: 5.5, 5.6, 6.3, 4.6, 5.3, 5.0, 6.2, 5.8, 5.1, 5.2, 5.9.

Model B: 3.8, 4.3, 4.2, 4.0, 4.9, 4.5, 5.2, 4.8, 4.5, 3.9, 3.7, 4.6.

Assume the populations to be approximately normal with equal variances. Can you think that the average running time of the calculator model A is longer than that of Model B ($\alpha=0.01$)?

Solution:

Let X be the running time of model A, $X\sim N(\mu_1,\sigma^2)$, and let Y be the running time of model B, $Y\sim N(\mu_2,\sigma^2)$. The hypotheses to be tested at the significance level $\alpha=0.01$ are

$$H_0:\mu_1-\mu_2=0, H_1:\mu_1-\mu_2>0.$$

The test statistic is

$$T=\frac{\overline{X}-\overline{Y}}{S_p\sqrt{\frac{1}{n_1}+\frac{1}{n_2}}}.$$

$n_1=11$, $n_2=12$, $\overline{x}=\frac{5.5+5.6+\cdots+5.9}{11}=5.5$, $\overline{y}=\frac{3.8+4.3+\cdots+4.6}{12}=4.367$, $s_p=0.495$, $\alpha=0.01$, $t_\alpha(n_1+n_2-2)=t_{0.01}(21)=2.517$.

The rejection region is

$$t>t_{0.01}(21)=2.517.$$

The value of $t=\dfrac{5.5-4.37}{0.495\times\sqrt{\dfrac{1}{11}+\dfrac{1}{12}}}=5.484>2.517$. The sample data fall in the rejection region, so we reject the null hypothesis. The sample data indicate that the average running time of the calculator of model A is longer than that of model B under the significance level $\alpha=0.01$.

Problem 7.18 Select two samples independently from two coal mines. The ash rates are measured(unit:%).

A:24.3,20.3,23.7,21.3,17.4.

B:18.2,16.9,20.2,16.7.

Assume the ash contents obey normal distributions with equal variances. Are there any significant differences in the average ash content of the two coal mines? ($\alpha=0.05$)

 **Solution**:

Let X be the ash rate for the first coal mine, $X\sim N(\mu_1,\sigma^2)$, and let Y be the ash rate for the second coal mine, $Y\sim N(\mu_2,\sigma^2)$.

The hypotheses to be tested at the significance level $\alpha=0.05$ are

$$H_0:\mu_1-\mu_2=0,H_1:\mu_1-\mu_2\neq 0.$$

The test statistic is

$$T=\frac{\overline{X}-\overline{Y}}{S_p\sqrt{\dfrac{1}{n_1}+\dfrac{1}{n_2}}}.$$

$n_1=5$, $n_2=4$, $\bar{x}=\dfrac{24.3+\cdots+17.4}{5}=21.4$, $\bar{y}=\dfrac{18.2+\cdots+16.7}{4}=18$, $s_p=2.351$, $\alpha=0.05$, $t_{\alpha/2}(n_1+n_2-2)=t_{0.025}(7)=2.364$. The rejection region is

$$|t|>t_{0.025}(7)=2.364.$$

The value of $|t|=\left|\dfrac{21.4-18}{2.351\sqrt{\dfrac{1}{5}+\dfrac{1}{4}}}\right|=2.156<2.364.$

The sample data fall outside the rejection region, so we fail to reject the null hypothesis. The sample data do not rule out the possibility that there are no significant differences in the average ash content of the two coal mines.

Problem 7.19 An investigator predicts that dog owners in the country spend more time walking their dogs than do dog owners in the city. The investigator gets a sample of 21 country owners and 23 city owners. The mean number of hours per week that city owners spend walking their dogs is 10.0. The standard deviation of hours spent walking the dog by city owners is 3.0. The mean number of hours country owners spent walking theirs dogs per week is 15.0. The standard deviation of the number of hours spent walking the dog by owners in the country is 4.0. Do dog owners in the country spend more time walking their dogs than do dog owners in the city? Assume the populations are normal with equal variances. ($\alpha=0.05$)

 **Solution**:

Let X be the number of hours spent walking the dog by owners in the country, $X\sim N(\mu_1,\sigma^2)$, and let Y be the number of hours spent walking the dog by owners in the city, $Y\sim N(\mu_2,\sigma^2)$.

The hypotheses to be tested at the significance level $\alpha=0.05$ are

$$H_0:\mu_1-\mu_2=0, H_1:\mu_1-\mu_2>0.$$

The test statistic is

$$T=\frac{\overline{X}-\overline{Y}}{S_p\sqrt{\frac{1}{n_1}+\frac{1}{n_2}}}.$$

$n_1=21, n_2=23, \overline{x}=15, \overline{y}=10, s_1=4, s_2=3, s_p=3.512, \alpha=0.05, t_\alpha(n_1+n_2-2)=t_{0.05}(42)=$ 1.68

The rejection region is

$$t>t_{0.05}(42)=1.68.$$

The value of $t=\frac{15-10}{3.512\times\sqrt{\frac{1}{21}+\frac{1}{23}}}=4.717>1.68$. The sample data fall in the rejection region, so we reject the null hypothesis. The sample data indicate that dog owners in the country spend more time walking their dogs than dog owners in the city do under the significance level $\alpha=0.05$.

Problem 7.20 From the two veins of a zinc mine, select two samples of size 9 and 8, respectively, the average zinc contents (unit: %) and the sample variances are as follows:

East Branch: $\overline{x}_1=0.230, s_1^2=0.1337$.

West Branch: $\overline{x}_2=0.269, s_2^2=0.1736$.

Assume the contents of zinc of two veins are normal with equal variances. Are the average values of zinc content of two veins the same? ($\alpha=0.05$)

 **Solution**:

Let X be the zinc content from the East branch, $X\sim N(\mu_1,\sigma^2)$, and let Y be the zinc content from the West branch, $Y\sim N(\mu_2,\sigma^2)$.

The hypotheses to be tested at the significance level $\alpha=0.05$ are

$$H_0:\mu_1-\mu_2=0, H_1:\mu_1-\mu_2\neq 0.$$

The test statistic is

$$T=\frac{\overline{X}_1-\overline{X}_2}{S_p\sqrt{\frac{1}{n_1}+\frac{1}{n_2}}}.$$

$n_1=9, n_2=8, \bar{x}_1=0.230, \bar{x}_2=0.269, S_p=0.39, \alpha=0.05, t_{\alpha/2}(n_1+n_2-2)=t_{0.025}(15)=2.131$. The rejection region is

$$|t|>t_{0.025}(15)=2.131.$$

The value of $|t|=\left|\frac{0.230-0.269}{0.39\sqrt{\frac{1}{9}+\frac{1}{8}}}\right|=0.205\ 8<2.131.$

The sample data fall outside the rejection region, so we fail to reject the null hypothesis. The sample data do not rule out the possibility that the average values of zinc content of two veins are the same.

Problem 7.21 A factory produces copper wire. The variance of the tensile force must not exceed 16(kg^2). From the production of copper wire, randomly select 9 wires, the tensile forces(unit:kg) are

$$289, 286, 285, 284, 286, 285, 286, 298, 292.$$

Assume that the tensile force obeys the normal distribution of $X\sim N(\mu,\sigma^2)$. Does the tensile force of the copper wire produced conform to the standard? ($\alpha=0.05$)

 **Solution 1:**

We want to test the hypotheses:

$$H_0:\sigma^2\leqslant 16, H_1:\sigma^2>16.$$

The test statistic under H_0 is

$$\chi^2=\frac{(n-1)S^2}{\sigma_0^2}\sim\chi^2(n-1).$$

$$n=9, s^2=\frac{1}{n-1}\sum_{i=1}^{n}(x_i-\bar{x})^2=20.361, \sigma_0=4, \chi_\alpha^2(n-1)=\chi_{0.05}^2(8)=15.51.$$

The rejection region is

$$W=\left\{(x_1,\cdots,x_n)\left|\frac{(n-1)s^2}{\sigma_0^2}\geqslant\chi_\alpha^2(n-1)\right.\right\}=\left\{(x_1,\cdots,x_n)\left|\frac{(n-1)s^2}{\sigma_0^2}\geqslant 15.51.\right.\right\}.$$

The value of the test statistic is $\chi^2=\frac{(9-1)\times 20.361}{16}=10.18<15.51$. The sample data fall outside the rejection region. That is, we fail to reject the null hypothesis. The sample data do not rule out the possibility that the tensile force of the copper wire produced conforms to the standard.

Solution 2:

We can also adopt the p-value approach to the hypothesis testing. In this case, the p-value is the probability that we would observe a $\chi^2(8)$ random variable more extreme than $\frac{(9-1)\times 20.361}{16}=10.18$, as shown in Figure 7.5.

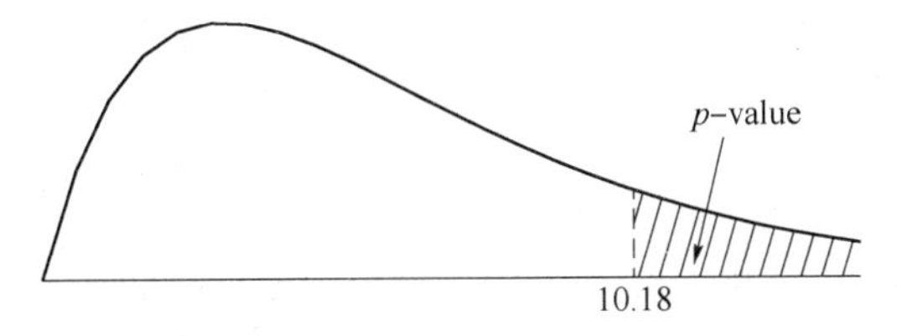

Figure 7.5

$$p=P(\chi^2(8)\geqslant 10.18)=0.2526>0.05=\alpha.$$

We do not reject the null hypothesis. The sample data do not rule out the possibility that the tensile force of the copper wire produced conforms to the standard. The p-value approach yields the same conclusion.

Problem 7.22 Select 10 fuses from a batch of fuses. The times needed to melt after the strong current (unit: second) are

$$42, 65, 75, 78, 59, 71, 57, 68, 54, 55.$$

Assume the time required to melt obeys the normal distribution.

(1) Is the average melting time not less than 65? ($\alpha=0.05$)

(2) Does the variance of the melting time not exceed 80? ($\alpha=0.05$)

Solution:

(1) Let X be the melting time, $X\sim N(\mu,\sigma^2)$. We want to test the hypotheses:

$$H_0:\mu\geqslant 65, H_1:\mu<65.$$

Since the variance of the population σ^2 is unknown, we use the test statistic

$$T=\frac{\overline{X}-\mu_0}{S/\sqrt{n}}.$$

$n=10, \overline{x}=62.4, s=11.037, \alpha=0.05, t_\alpha(n-1)=t_{0.05}(9)=1.833$. The rejection region is

$$t=\frac{\overline{x}-\mu_0}{s/\sqrt{n}}<-t_{0.05}(9)=-1.833.$$

The value of $t=\frac{62.4-65}{11.037/\sqrt{10}}=-0.745>-1.833$. The sample data fall outside the rejection region. That is, we fail to reject the null hypothesis. The sample data do not rule out the possibility that the average melting time is not less than 65 seconds.

(2) We want to test the hypotheses:

$$H_0:\sigma^2\leqslant 80, H_1:\sigma^2>80.$$

The test statistic is

$$\chi^2=\frac{(n-1)S^2}{\sigma_0^2}.$$

$n=10, s^2=121.81, \alpha=0.05, \chi_\alpha^2(n-1)=\chi_{0.05}^2(9)=16.919$. The rejection region is

$$\chi^2>\chi_{0.05}^2(9)=16.919.$$

The value of $\chi^2=\frac{9\times 121.81}{80}=13.703<16.919$. The sample data fall outside the rejection region. That is, we fail to reject the null hypothesis. The sample data do not rule out the possibility that the variance of the melting time does not exceed 80 seconds.

Problem 7.23 Two lathes produce the same kind of bearings. The diameters of the bearings obey the normal distribution. Select 8 and 9 products separately from the two lathes. The diameters are measured. (unit: mm)

A Lathe: 15.0, 14.5, 15.2, 15.5, 14.8, 15.1, 15.2, 14.8.

B Lathe: 15.2, 15.0, 14.8, 15.2, 15.0, 15.0, 14.8, 15.1, 14.8.

Do the population variances of the bearings diameters produced by two lathes have a significant difference? ($\alpha=0.05$)

Solution 1:

Let the diameters of the bearings obey the normal distributions $N(\mu_1,\sigma_1^2)$ and $N(\mu_2,\sigma_2^2)$, respectively. We want to test the hypotheses:

$$H_0:\sigma_1^2=\sigma_2^2, H_1:\sigma_1^2\neq\sigma_2^2.$$

The test statistic under H_0 is

$$F=\frac{S_1^2}{S_2^2}.$$

At the significant level $\alpha=0.05$, the rejection region is

$$F\leqslant F_{0.975}(7,8)=0.204 \text{ or } F\geqslant F_{0.025}(7,8)=4.528.$$

The value of $F=\frac{s_1^2}{s_2^2}=\frac{0.0955}{0.0261}=3.659$, $0.204<3.659<4.528$, so we fail to reject the null hypothesis H_0. The sample data do not rule out the possibility that the population variances of the bearings diameters produced by two lathes are same.

Solution 2:

In this case, the p-value is given as follows:

$$p=2\min\{P[F(7,8)\geqslant 3.659], P[F(7,8)\leqslant 3.659]\}=0.089>0.05=\alpha.$$

So we fail to reject the null hypothesis H_0. The p-value approach yields the same conclusion.

Problem 7.24 A study was performed on patients with pituitary adenomas. The standard deviation of the weights of 12 patients with pituitary adenomas was 21.4 kg. A control group of 5 patients without pituitary adenomas had a standard deviation of the weights of 12.4 kg. We wish to know if the weights of the patients with pituitary adenomas are more variable than the weights of the control group. Conduct a statistical test. ($\alpha=0.05$)

Solution 1:

Let σ_1^2 and σ_2^2 be the variances of the patients with pituitary adenomas and the control group, respectively, we want to test the hypotheses:

$$H_0:\sigma_1^2\leqslant\sigma_2^2, H_1:\sigma_1^2>\sigma_2^2.$$

The test statistic under H_0 is: $F=\dfrac{S_1^2}{S_2^2}\sim F(m-1,n-1)$.

$$m=12, n=5, \alpha=0.05, F_\alpha(m-1,n-1)=F_{0.05}(11,4)=5.936.$$

The rejection region is

$$W=\left\{(x_1,\cdots,x_n)\,\middle|\,\frac{s_1^2}{s_2^2}\geqslant F_\alpha(m-1,n-1)\right\}=\left\{(x_1,\cdots,x_n)\,\middle|\,\frac{s_1^2}{s_2^2}\geqslant 5.936\right\}.$$

We have $s_1^2=21.4^2, s_2^2=12.4^2$, the value of the test statistic is $F=\dfrac{s_1^2}{s_2^2}=2.978$. It falls outside the rejection region, so we fail to reject the null hypothesis. The weights of the patients with pituitary adenomas are not more variable than the weights of the control group.

Solution 2:

In this case, the p-value is the probability that we would observe an $F(11,4)$ random variable more extreme than 2.978, as shown in Figure 7.6.

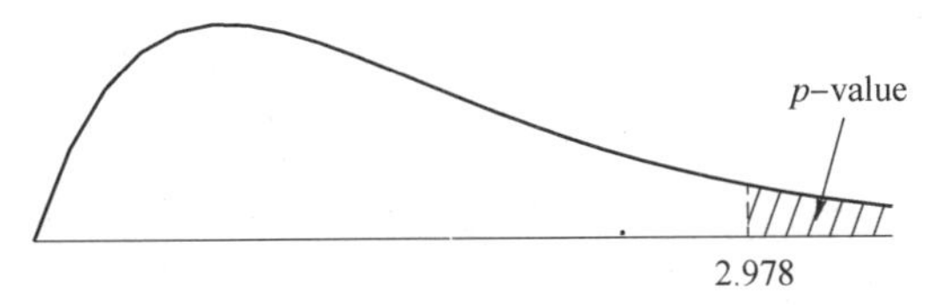

Figure 7.6

$$p=P[F(11,4)\geqslant 2.978]=0.151\ 7>0.05=\alpha.$$

So we fail to reject the null hypothesis. The weights of the patients with pituitary adenomas are not more variable than the weights of the control group. The p-value approach yields the same conclusion.

Problem 7.25 Select two samples independently from two batches of electronic devices. The resistance values(unit:Ω) are as follows:

A batch:0.140,0.138,0.143,0.142,0.144,0.137.

B batch:0.135,0.140,0.142,0.136,0.138,0.140.

Assume the resistance values of two batches are $N(\mu_1,\sigma_1^2)$ and $N(\mu_2,\sigma_2^2)$, respectively.

(1) Test whether the variances of two populations are equal ($\alpha=0.05$);

(2) Test whether the mean values of two populations are equal ($\alpha=0.05$).

 **Solution**:

(1) The hypotheses to be tested at the significance level $\alpha=0.05$ are

$$H_0:\sigma_1^2=\sigma_2^2, H_1:\sigma_1^2\neq\sigma_2^2.$$

The test statistic under H_0 is

$$F=\frac{S_1^2}{S_2^2}.$$

The rejection region is

$$F\leqslant F_{0.975}(5,5)=0.139 \text{ or } F\geqslant F_{0.025}(5,5)=7.146.$$

The value of $F=\frac{s_1^2}{s_2^2}=\frac{7.866\times10^{-6}}{7.1\times10^{-6}}=1.1079$, $0.139<1.1079<7.146$. We do not reject H_0. The sample data do not rule out the possibility that the variances of two populations are equal.

(2) The hypotheses to be tested at the significance level $\alpha=0.05$ are

$$H_0:\mu_1-\mu_2=0, H_1:\mu_1-\mu_2\neq0.$$

The test statistic is

$$T=\frac{\overline{X}-\overline{Y}}{S_p\sqrt{\frac{1}{n_1}+\frac{1}{n_2}}}.$$

$n_1=6, n_2=6, \bar{x}=0.1407, \bar{y}=0.1385, s_p=0.00273, \alpha=0.05, t_{\alpha/2}(n_1+n_2-2)=t_{0.025}(10)=2.228$. The rejection region is

$$|t|>t_{0.025}(10)=2.228.$$

The value of $|t|=\left|\frac{0.1407-0.1385}{0.00273\sqrt{\frac{1}{6}+\frac{1}{6}}}\right|=1.374<2.228$. We do not reject H_0. The sample data do not rule out the possibility that the mean values of two populations are equal.

Problem 7.26 Suppose that we roll a die 30 times and observe the following Table 7.1 showing the number of times each face ends up on top. Is the die fair? ($\alpha=0.05$)

Table 7.1

Face	1	2	3	4	5	6
Count	3	7	5	10	2	3

Solution 1:

Let p_i be the probability that face i ends up on top, $i=1,\cdots,6$. We want to test

$$H_0: p_1=\cdots=p_6=\frac{1}{6}.$$

H_1: at least one p_i is not as specified in H_0.

The test statistic value is

$$\chi^2=\sum_{i=1}^{6}\frac{(n_i-np_i)^2}{np_i}$$

$$=\frac{\left(3-30\times\frac{1}{6}\right)^2}{30\times\frac{1}{6}}+\frac{\left(7-30\times\frac{1}{6}\right)^2}{30\times\frac{1}{6}}+\frac{\left(5-30\times\frac{1}{6}\right)^2}{30\times\frac{1}{6}}+\frac{\left(10-30\times\frac{1}{6}\right)^2}{30\times\frac{1}{6}}+\frac{\left(2-30\times\frac{1}{6}\right)^2}{30\times\frac{1}{6}}+\frac{\left(3-30\times\frac{1}{6}\right)^2}{30\times\frac{1}{6}}$$

$$=9.2.$$

At the significance level $\alpha=0.05$, $\chi^2_{0.05}(5)=11.070\ 5$, $9.2<11.070\ 5$, we fail not reject the null hypothesis H_0. The die is fair.

Solution 2:

In this case, the p-value is the probability that we would observe a $\chi^2(5)$ random variable more extreme than 9.2, as shown in Figure 7.7.

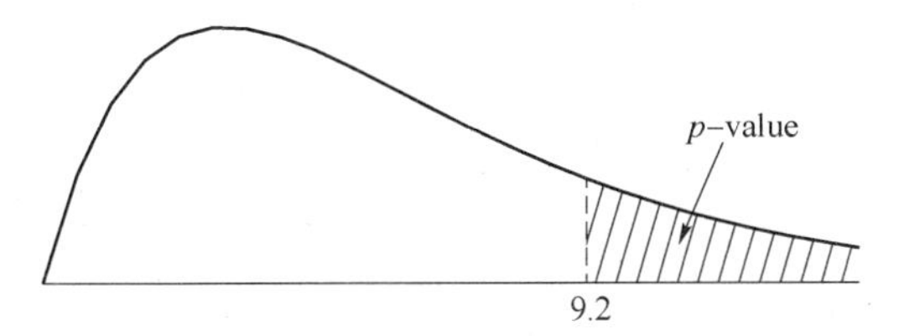

Figure 7.7

$$p=P[\chi^2(5)\geqslant 9.2]=0.101\ 3>0.05=\alpha.$$

We fail to reject the null hypothesis in favor of the alternative hypothesis. The die is fair. The p-value approach yields the same conclusion.

Problem 7.27 Tall cut-leaf tomatoes were crossed with dwarf potato-leaf tomatoes, and $n = 1\ 611$ offsprings were classified by their phenotypes, as shown in Table 7.2.

Table 7.2

Phenotypes	tall cut-leaf	tall potato-leaf	dwarf cut-leaf	dwarf potato-leaf
Count	926	288	293	104

Genetic theory says that the four phenotypes should occur with relative frequencies 9 : 3 : 3 : 1, and thus are not all equally as likely to be observed. Do the observed data support this theory? ($\alpha=0.05$)

 **Solution 1:**

Let p_i be the probability that the i-th phenotype appears, $i=1,\cdots,4$. We want to test

$$H_0: p_1=\frac{9}{16}, p_2=p_3=\frac{3}{16}, p_4=\frac{1}{16}.$$

$$H_1: \text{at least one } p_i \text{ is not as specified in } H_0.$$

The test statistic value is

$$\chi^2 = \sum_{i=1}^{4} \frac{(n_i - np_i)^2}{np_i}$$

$$=\frac{\left(926-1\ 611\times\frac{9}{16}\right)^2}{1\ 611\times\frac{9}{16}}+\frac{\left(288-1\ 611\times\frac{3}{16}\right)^2}{1\ 611\times\frac{3}{16}}+\frac{\left(293-1\ 611\times\frac{3}{16}\right)^2}{1\ 611\times\frac{3}{16}}+\frac{\left(104-1\ 611\times\frac{1}{16}\right)^2}{1\ 611\times\frac{1}{16}}$$

$$=1.47.$$

At the significance level $\alpha=0.05, \chi^2_{0.05}(3) = 7.815, 1.47<7.815$, we fail not reject the null hypothesis H_0. The observed data support this genetic theory.

 **Solution 2:**

In this case, the p-value is the probability that we would observe a $\chi^2(3)$ random variable more extreme than 1.47, as shown in Figure 7.8.

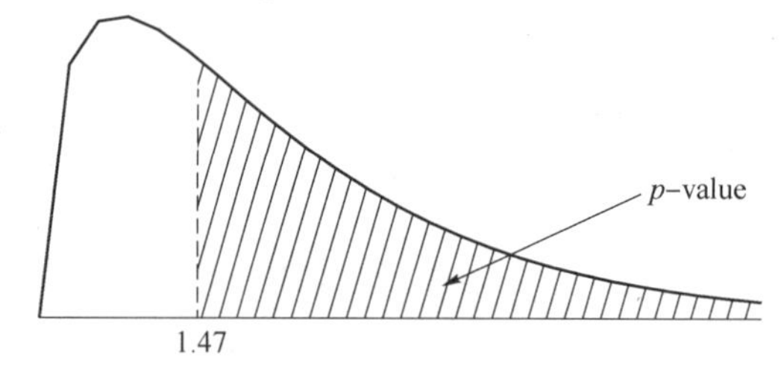

Figure 7.8

$$p=P[\chi^2(3)\geqslant 1.47]=0.689\ 2>0.05=\alpha.$$

We fail to reject the null hypothesis in favor of the alternative hypothesis. The observed data support this genetic theory. The p-value approach yields the same conclusion.

Problem 7.28 There are 100 batches of products, each containing 10 products. Some are defective. See Table 7.3. Does the number of defective products in a batch follow a binomial distribution? ($\alpha=0.05$)

Table 7.3

The number of defective products	0	1	2	3	4	5	6	7	8	9	10
Count	35	40	18	5	1	1	0	0	0	0	0

 **Solution**:

Let X be the number of defectives in a batch. We want to test

$$H_0: X\sim \mathrm{B}(10,p),H_1: X \text{ does not follow } \mathrm{B}(10,p).$$

Under the condition that H_0 is true, the maximum likelihood estimate of p is

$$\hat{p}=\frac{1}{100\times 10}\times(0\times 35+1\times 40+\cdots+10\times 0)=0.1.$$

The PF of X under H_0 is

$$P(X=i)=\mathrm{C}_{10}^{i}\times 0.1^{i}\times 0.9^{10-i},i=0,1,2,\cdots,10.$$

We combine the frequencies with the number of defective products greater than or equal to 3 into one group, and the corresponding probabilities are

$$\hat{p}_0=0.348\ 6,\hat{p}_1=0.387\ 4,\hat{p}_2=0.193\ 7,\hat{p}_3=\sum_{i=3}^{10}\mathrm{C}_{10}^{i}\times 0.1^{i}\times 0.9^{10-i}=0.070\ 2.$$

At the significant level $\alpha=0.05,\chi^2_{0.05}(k-r-1)=\chi^2_{0.05}(2)=5.991$, where r is the number of unknown parameters, k is the number of groups. The value of

$$\chi^2=\frac{(35-100\times 0.348\ 6)^2}{100\times 0.348\ 6}+\frac{(40-100\times 0.387\ 4)^2}{100\times 0.387\ 4}+\frac{(18-100\times 0.183\ 7)^2}{100\times 0.183\ 7}+\frac{(7-100\times 0.070\ 2)^2}{100\times 0.070\ 2}$$

$$=0.138<5.991.$$

We fail to reject the null hypothesis in favor of the alternative hypothesis. The number of defective products in a batch follows a binomial distribution.

Problem 7.29 A study investigates the association between smoking and asthma among adults observed in a community health clinic. The results obtained from classifying 150 individuals are shown in Table 7.4. Does cigarette smoking affect asthma symptoms? ($\alpha=0.05$)

Table 7.4

Symptoms of Asthma	Ever Smoked Cigarettes		Total
	Yes	No	
Yes	20	30	50
No	22	78	100
Total	42	108	150

Solution 1:

H_0: Cigarette smoking is independent of asthma symptoms.

H_1: Cigarette smoking affects asthma symptoms.

The test statistic value is

$$\chi^2 = \frac{\left(20-\frac{50\times42}{150}\right)^2}{\frac{50\times42}{150}} + \frac{\left(30-\frac{50\times108}{150}\right)^2}{\frac{50\times108}{150}} + \frac{\left(22-\frac{100\times42}{150}\right)^2}{\frac{100\times42}{150}} + \frac{\left(78-\frac{100\times108}{150}\right)^2}{\frac{100\times108}{150}}$$

$$=5.357.$$

Using the 0.05 level of significance, the critical value of χ^2 with $(2-1)\times(2-1)=1$ degree of freedom is 3.841. The decision rule is "reject H_0 if $\chi^2>3.841$; otherwise do not reject H_0". Since $5.357>3.841$, we reject the null hypothesis H_0. There is sufficient evidence at the 0.05 level to conclude that cigarette smoking affects asthma symptoms.

Solution 2:

In this case, the p-value is the probability that we would observe a $\chi^2(1)$ random variable more extreme than 5.357, as shown in Figure 7.9.

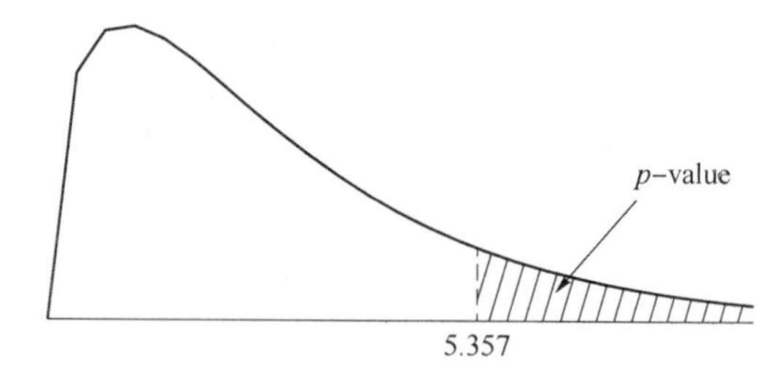

Figure 7.9

$$p=P[\chi^2(1)\geqslant 5.357]=0.02<0.05=\alpha.$$

We reject the null hypothesis. There is sufficient evidence at the 0.05 level to conclude that cigarette smoking affects asthma symptoms. The p-value approach yields the same conclusion.

Problem 7.30 Should the proportion of elementary schools students' computer courses be improved? To study this problem, a survey has been conducted and the results are reported in Table 7.5. Does the factor of age affect the answer to this question? $(\alpha=0.05)$

Table 7.5

Age	Agree	Disagree	NA
over 55	32	28	14
36~55	44	21	17
15~35	47	12	13

 **Solution 1:**

H_0: the factor of age is independent of the answer to this question, that is, $p_{ij}=p_{i\cdot}\times p_{\cdot j}, i=1, 2, 3.$

H_1: the factor of age affects the answer to this question.

The test statistic is

$$\chi^2=\sum_{i=1}^{3}\sum_{j=1}^{3}\frac{(n_{ij}-n\hat{p}_{ij})^2}{n\hat{p}_{ij}}\sim\chi^2(4).$$

Using the significant level $\alpha=0.05$, the critical value of χ^2 with $(3-1)\times(3-1)=4$ degrees of freedom is $\chi^2_{0.05}(4)=9.487\ 7$. The decision rule is "reject H_0 if $\chi^2>9.487\ 7$; otherwise do not reject H_0".

The expected counts $\frac{n_{i\cdot}\times n_{\cdot j}}{n}$ are reported in Table 7.6. The test statistic value is

$$\chi^2=\frac{(32-39.921)^2}{39.921}+\cdots+\frac{(13-13.895)^2}{13.895}=9.613\ 2>9.487\ 7.$$

Table 7.6

Age	Agree	Disagree	NA	Total
over 55	39.921	19.798	14.281	74
36~55	44.237	21.939	15.825	82
15~35	38.842	19.263	13.895	72
Total	123	61	44	228

We reject the null hypothesis H_0. There is sufficient evidence at the 0.05 level to conclude that the factor of age affects the answer to this question.

 **Solution 2:**

In this case, the p-value is the probability that we would observe a $\chi^2(4)$ random variable more extreme than 9.613 2, as shown in Figure 7.10.

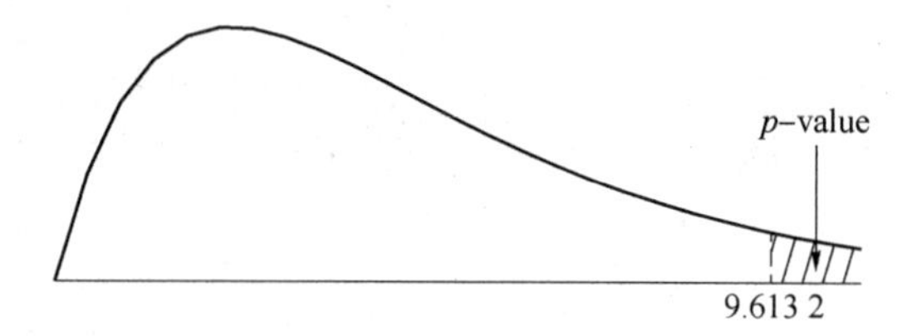

Figure 7.10

$$p=P[\chi^2(4)\geqslant 9.613\ 2]=0.047\ 4<0.05=\alpha.$$

We reject the null hypothesis. There is sufficient evidence at the 0.05 level to conclude that the factor of age affects the answer to this question. The p - value approach yields the same conclusion.

Chapter 8 Linear Regression

Summary of Knowledge

Exercise Solutions

Problem 8.1 To investigate the problem of the solubility of a substance in water, the data of dissolution mass and temperature are obtained and shown below.

Temperature x_i(℃)	0	4	10	15	21	29	36	51	68
Dissolution mass Y_i(mg)	66.7	71.0	76.3	80.6	85.7	92.9	99.4	113.6	125.1

Assume Y obeys the linear model

$$Y=a+bx+\varepsilon,\varepsilon\sim N(0,\sigma^2).$$

Determine the estimates of a,b, and σ^2.

Solution:

$$n=9,\ \bar{x}=26,\bar{y}=90.144,\sum_{i=1}^{9}x_i=234,\sum_{i=1}^{9}y_i=811.3$$

$$\sum_{i=1}^{9}x_i^2=10\ 144,\sum_{i=1}^{9}y_i^2=76\ 218.17,\sum_{i=1}^{9}x_i y_i=24\ 628.6$$

Thus,

$$\hat{b}=\frac{S_{xy}}{S_{xx}}=\frac{24\ 628.6-\frac{1}{9}\times234\times811.3}{10\ 144-\frac{1}{9}\times234^2}=0.87$$

$$\hat{a}=\bar{y}-\hat{b}\bar{x}=90.144-0.87\times26=67.524.$$

$$\hat{\sigma}^2=\frac{\text{SSE}}{n-2}=\frac{S_{yy}-\hat{b}S_{xy}}{n-2}$$

$$=\frac{76\ 218.17-\frac{1}{9}\times 811.3^2-0.87\times\left(24\ 628.6-\frac{1}{9}\times 234\times 811.3\right)}{9-2}$$

$=0.92.$

Problem 8.2 The relationship between the reproduction quantity and the month of a substance is as follows:

Month x_i	2	4	6	8	10
Reproduction quantity Y_i	66	120	210	270	320

Suppose that the linear model is obeyed. $E(Y)=\mu(x)=\beta_0+\beta_1 x$.

(1) Find the estimates of β_0 and β_1.

(2) Test whether β_1 is equal to zero or not. ($\alpha=0.05$)

 **Solution:**

(1)

$$n=5,\ \bar{x}=6, \bar{y}=197.2,\ \sum_{i=1}^{5} x_i=30,\ \sum_{i=1}^{5} y_i=986$$

$$\sum_{i=1}^{5} x_i^2=220,\ \sum_{i=1}^{5} y_i^2=238\ 156,\ \sum_{i=1}^{5} x_i y_i=7\ 232$$

Thus, $\hat{\beta}_1=\dfrac{S_{xy}}{S_{xx}}=\dfrac{7\ 232-\frac{1}{5}\times 30\times 986}{220-\frac{1}{5}\times 30^2}=32.9, \hat{\beta}_0=\bar{y}-\hat{\beta}_1\bar{x}=197.2-32.9\times 6=-0.2.$

(2) Method 1:

At the significant level $\alpha=0.05$, we want to test

$$H_0:\beta_1=0, H_1:\beta_1\neq 0.$$

The test statistic is

$$T=\frac{\hat{\beta}_1}{\hat{\sigma}}\sqrt{S_{xx}}=\frac{\hat{\beta}_1\sqrt{S_{xx}}}{\sqrt{\frac{\text{SSE}}{n-2}}}\sim t(n-2).$$

The rejection region is

$$|t|\geqslant t_{\alpha/2}(n-2)=t_{0.025}(3)=3.182.$$

Since $\hat{\sigma}=\sqrt{\dfrac{\text{SSE}}{n-2}}=\sqrt{\dfrac{S_{yy}-\hat{\beta}_1 S_{xy}}{n-2}}=\sqrt{\dfrac{420.4}{3}}=11.837$, $|t|=\left|\dfrac{32.9}{11.837}\times\sqrt{40}\right|=17.57>3.182$, the sample data fall in the rejection region, we reject H_0. β_1 is not equal to zero.

In this case, the p-value is

$$p=P[\ |t(3)|\geqslant 17.57]=0.000\ 4<0.05.$$

The p-value approach yields the same conclusion.

Method 2:

$$SST = S_{yy} = 43\ 716.8, df = 5-1 = 4$$

$$SSR = \hat{\beta}_1^2 S_{xx} = 32.9^2 \times 40 = 43\ 296.4, df = 1$$

$$SSE = SST - SSR = 420.4, df = 3$$

$$F = \frac{\frac{43\ 296.4}{1}}{\frac{420.4}{3}} = 308.9.$$

The calculations are reported in Table 8.1.

Table 8.1

Source	Sum of squares	Degrees of Freedom	Mean Square	F-value	p-value
Model	SSR = 43 296.4	1	43 296.4	308.9	0.000 4
Error	SSE = 420.4	3	140.13	—	—
Total	SST = 43 716.8	4	—	—	—

At the level $\alpha = 0.05$, $308.9 > F_{0.05}(1,3) = 10.127\ 9$, we reject H_0. The regression model is significant.

In this case, the p-value is the probability that we would observe an $F(1,3)$ random variable more extreme than 308.9,

$$p = P[F(1,3) \geqslant 308.9] = 0.000\ 4 < 0.05.$$

The p-value approach yields the same conclusion.

Problem 8.3 The strength Y (unit: MPa) of a synthetic material and its tensile multiples x have the following relationship:

x_i	2.0	2.5	2.7	3.5	4.0	4.5	5.2	6.3	7.1	8.0	9.0	10.0
Y_i	1.3	2.5	2.5	2.7	3.5	4.2	5.0	6.4	6.3	7.0	8.0	8.1

(1) Find the linear regression equation.

(2) Test the significance of the regression line. ($\alpha = 0.05$)

(3) Find a 95% prediction interval of Y_0 when $x_0 = 6$.

Solution:

(1)

$$n = 12,\ \bar{x} = 5.4, \bar{y} = 4.79,\ \sum_{i=1}^{12} x_i = 64.8,\ \sum_{i=1}^{12} y_i = 57.5$$

$$\sum_{i=1}^{12} x_i^2 = 428.18, \sum_{i=1}^{12} y_i^2 = 335.63, \sum_{i=1}^{12} x_i y_i = 378.$$

Therefore, $\hat{\beta}_1 = \frac{S_{xy}}{S_{xx}} = \frac{378-\frac{1}{12}\times 64.8\times 57.5}{428.18-\frac{1}{12}\times 64.8^2} = 0.862\ 5$, $\hat{\beta}_0 = \bar{y} - \hat{\beta}_1 \bar{x} = 4.79 - 0.862\ 5\times 5.4 = 0.132\ 5$, the regression equation is

$$\hat{y} = 0.132\ 5 + 0.862\ 5x.$$

(2) Method 1:

At the significant level $\alpha=0.05$, we want to test

$$H_0: \beta_1 = 0, H_1: \beta_1 \neq 0.$$

The test statistic is

$$T = \frac{\hat{\beta}_1}{\hat{\sigma}}\sqrt{S_{xx}} \sim t(n-2).$$

The rejection region is

$$|t| \geqslant t_{\alpha/2}(n-2) = t_{0.025}(10) = 2.228.$$

$\hat{\sigma} = \sqrt{\frac{SSE}{n-2}} = \sqrt{\frac{1.88}{10}} = 0.433\ 6$, $|t| = \left|\frac{0.862\ 5}{0.433\ 6}\times\sqrt{78.26}\right| = 17.6 > 2.228$. Since the sample data fall in the rejection region, we reject H_0. β_1 is not zero. The regression line is significant.

In this case, the p-value is

$$p = P[\,|t(10)| \geqslant 17.6] = 7.5\times 10^{-9} < 0.05.$$

The p-value approach yields the same conclusion.

Method 2:

$$SST = S_{yy} = 335.63 - \frac{1}{12}\times 57.5^2 = 60.11, df = 12-1 = 11$$

$$SSR = \hat{\beta}_1^2 S_{xx} = 0.862\ 5^2 \times \left(428.18 - \frac{1}{12}\times 64.8^2\right) = 58.23, df = 1$$

$$SSE = SST - SSR = 1.88, df = 10$$

$$F = \frac{\frac{58.23}{1}}{\frac{1.88}{10}} = 309.73.$$

The calculations are reported in Table 8.2.

Table 8.2

Source	Sum of squares	Degrees of Freedom	Mean Square	F-value	p-value
Model	SSR = 58.23	1	58.23	309.73	7.5×10^{-9}
Error	SSE = 1.88	10	0.188	—	—
Total	SST = 60.11	11	—	—	—

At the level $\alpha=0.05, 309.73>F_{0.05}(1,10)=4.965$, we reject H_0. The regression model is significant.

In this case, the p-value is the probability that we would observe an $F(1,10)$ random variable more extreme than 309.73,

$$p=P[F(1,10)\geqslant 309.74]=7.5\times 10^{-9}<0.05.$$

The p-value approach yields the same conclusion.

(3) When $x_0=6$, the predicted value of y_0 is $\hat{y}_0=\hat{\beta}_0+\hat{\beta}_1 x_0=0.132\,5+0.862\,5\times 6=5.307\,5$, and the 95% prediction interval is

$$\left(\hat{y}_0 \pm t_{\alpha/2}(n-2)\hat{\sigma}\sqrt{1+\frac{1}{n}+\frac{(x_0-\bar{x})^2}{S_{xx}}}\right)$$

$$=\left(5.307\,5\pm 2.228\times 0.433\,6\sqrt{1+\frac{1}{12}+\frac{(6-5.4)^2}{78.26}}\right)$$

$$=(4.299\,9, 6.315\,1).$$

See Figure 8.1.

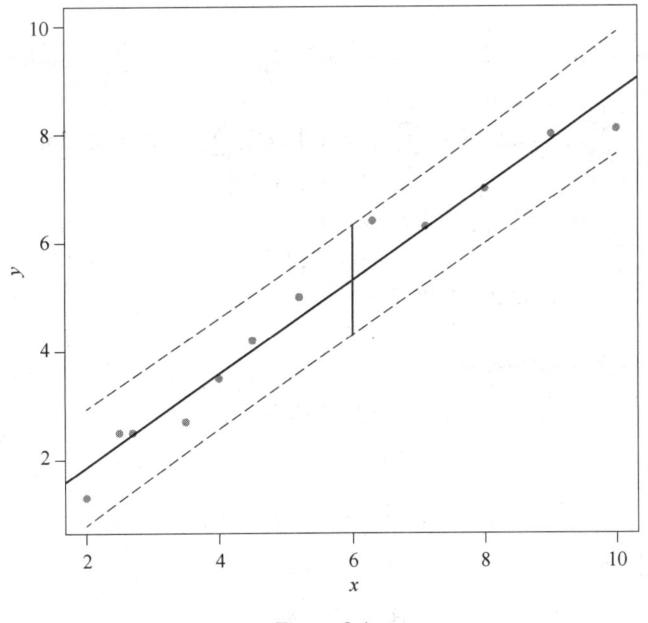

Figure 8.1

Problem 8.4 A statistics instructor at a university would like to examine the relationship (if any) between the number of optional homework problems students do during the semester and their final course grade. She randomly selects 12 students for study and asks them to keep track of the number of these problems completed during the course of the semester. The data are reported in Table 8.3 below:

Table 8.3

Problems	51	58	62	65	68	76	77	78	78	84	85	91
Grade	62	68	66	66	67	72	73	72	78	73	76	75

(1) For this setting identify the response variable and the predictor variable.

(2) Find the regression equation.

(3) Conduct a hypothesis test for the significance of the linear equation.

(4) Use the regression equation to predict a student's final course grade if 70 optional homework assignments are done.

 **Solution**:

(1) The response variable y is the final course grade. The predictor variable x is the number of optional homework problems students do during the semester.

(2)

$$n = 12,\ \bar{x} = 72.75,\ \bar{y} = 70.667,\ \sum_{i=1}^{12} x_i = 873,\ \sum_{i=1}^{12} y_i = 848$$

$$\sum_{i=1}^{12} x_i^2 = 65\ 093,\ \sum_{i=1}^{12} y_i^2 = 60\ 180,\ \sum_{i=1}^{12} x_i y_i = 62\ 254$$

Therefore, $\hat{\beta}_1 = \dfrac{S_{xy}}{S_{xx}} = \dfrac{62\ 254 - \frac{1}{12}\times 873\times 848}{65\ 093 - \frac{1}{12}\times 873^2} = 0.355\ 2$, $\hat{\beta}_0 = \bar{y} - \hat{\beta}_1\bar{x} = 70.667 - 0.355\ 2\times 72.75 =$ 44.826 2, the regression equation is

$$\hat{y} = 44.826\ 2 + 0.355\ 2x.$$

(3) Method 1: At the significant level $\alpha = 0.05$, we want to test

$$H_0: \beta_1 = 0, H_1: \beta_1 \neq 0.$$

The test statistic is

$$T = \frac{\hat{\beta}_1}{\hat{\sigma}}\sqrt{S_{xx}} \sim t(n-2).$$

The rejection region is

$$|t| \geqslant t_{\alpha/2}(n-2) = t_{0.025}(10) = 2.228.$$

$$\hat{\sigma} = \sqrt{\frac{SSE}{n-2}} = \sqrt{\frac{55.038\ 9}{10}} = 2.346,\ |t| = \left|\frac{0.355\ 2}{2.346}\times\sqrt{1\ 582.25}\right| = 6.022\ 6 > 2.228.$$

Since the sample data fall in the rejection region, we reject H_0. β_1 is not zero. The regression line is significant.

In this case, the p-value is

$$p=P[\,|t(10)|\geqslant 6.022\,6]=1.28\times10^{-4}<0.05.$$

The p-value approach yields the same conclusion.

Method 2:

$$SST=S_{yy}=60\,180-\frac{1}{12}\times848^2=254.666\,7, df=12-1=11$$

$$SSR=\hat{\beta}_1^2S_{xx}=0.355\,2^2\times\left(65\,093-\frac{1}{12}\times873^2\right)=199.627\,8, df=1$$

$$SSE=SST-SSR=55.038\,9, df=10$$

$$F=\frac{\dfrac{199.627\,8}{1}}{\dfrac{55.038\,9}{10}}=36.27.$$

The calculations are reported in Table 8.4

Table 8.4

Source	Sum of squares	Degrees of Freedom	Mean Square	F-value	p-value
Model	SSR = 199.627 8	1	199.627 8	36.27	1.28×10^{-4}
Error	SSE = 55.038 9	10	5.503 89	—	—
Total	SST = 254.666 7	11	—	—	—

At the level $\alpha=0.05$, $36.27>F_{0.05}(1,10)=4.965$, we reject H_0. The regression model is significant.

In this case, the p-value is the probability that we would observe an $F(1,10)$ random variable more extreme than 36.27,

$$p=P[F(1,10)\geqslant 36.27]=1.28\times10^{-4}<0.05.$$

The p-value approach yields the same conclusion.

(4) When $x_0=70$, the predicted value of y_0 is $\hat{y}_0=\hat{\beta}_0+\hat{\beta}_1x_0=44.826\,2+0.355\,2\times70=69.69$.

Problem 8.5 The sales of a company (in million dollars) for each year are shown in table 8.5.

Table 8.5

x(year)	2005	2006	2007	2008	2009
y(sales)	12	19	29	37	45

(1) Find the least square regression line $y=\beta_0+\beta_1x$.

(2) Use the least square regression line to estimate the sales of the company in 2012.

Solution:

(1)

$$n=5,\ \bar{x}=2\ 007,\ \bar{y}=28.4,\ \sum_{i=1}^{5}x_i=10\ 035,\ \sum_{i=1}^{5}y_i=142$$

$$\sum_{i=1}^{5}x_i^2=20\ 140\ 255,\ \sum_{i=1}^{5}y_i^2=4\ 740,\ \sum_{i=1}^{5}x_iy_i=285\ 078$$

$$S_{xy}=285\ 078-\frac{1}{5}\times 10\ 035\times 142=84,\ S_{xx}=20\ 140\ 255-\frac{1}{5}\times 10\ 035^2=10$$

Therefore, $\hat{\beta}_1=\dfrac{S_{xy}}{S_{xx}}=\dfrac{84}{10}=8.4$, $\hat{\beta}_0=\bar{y}-\hat{\beta}_1\bar{x}=28.4-8.4\times 2\ 007=-16\ 830.4$, the regression equation is

$$\hat{y}=-16\ 830.4+8.4x.$$

(2) When $x_0=2\ 012$, the estimated sales of the company is $\hat{y}_0=\hat{\beta}_0+\hat{\beta}_1x_0=-16\ 830.4+8.4\times 2\ 012=70.4$.

Problem 8.6 A dietetics student wants to look at the relationship between calcium intake (mg/day) and knowledge score (out of 50) about calcium in sports science students. Table 8.6 shows the data she collected.

Table 8.6

Respondent	Knowledge score	Calcium intake	Respondent	Knowledge score	Calcium intake
1	10	450	11	38	940
2	42	1 050	12	25	733
3	38	900	13	48	985
4	15	525	14	28	763
5	22	710	15	22	583
6	32	854	16	45	850
7	40	800	17	18	798
8	14	493	18	24	754
9	26	730	19	30	805
10	32	894	20	43	1 085

(1) Find the least square regression line.

(2) What is the predicted calcium intake for a student with knowledge of calcium score equal to 30 (out of 50)?

Solution:

(1) Let x be the knowledge score and y be the calcium intake.

$$n=20,\ \bar{x}=29.6,\ \bar{y}=785.1,\ \sum_{i=1}^{20}x_i=592,\ \sum_{i=1}^{20}y_i=15\ 702$$

$$\sum_{i=1}^{20}x_i^2=19\ 852,\ \sum_{i=1}^{20}y_i^2=12\ 905\ 468,\ \sum_{i=1}^{20}x_i y_i=497\ 143$$

$$S_{xy}=497\ 143-\frac{1}{20}\times592\times15\ 702=32\ 363.8,\ S_{xx}=19\ 852-\frac{1}{20}\times592^2=2\ 328.8.$$

Therefore, $\hat{\beta}_1=\dfrac{S_{xy}}{S_{xx}}=\dfrac{32\ 363.8}{2\ 328.8}=13.897\ 2$, $\hat{\beta}_0=\bar{y}-\hat{\beta}_1\bar{x}=785.1-13.897\ 2\times29.6=373.742\ 9$,

the regression equation is

$$\hat{y}=373.742\ 9+13.897\ 2x.$$

(2) When $x_0=30$, the estimated sales of the company is $\hat{y}_0=\hat{\beta}_0+\hat{\beta}_1x_0=373.742\ 9+13.897\ 2\times30$ $=790.658\ 9$.

Problem 8.7 In animal studies, it is sometimes necessary to find out the relationship between the volume and the weight of a certain animal. Since the weight of an animal is relatively easy to measure, and the measurement of volume is more difficult, people want to use the weight of an animal to predict the volume. Table 8.7 reports the data of volumes and weights for 18 certain animals. Here, the weight x (unit: kg) is considered as an independent variable and the volume y (unit: dm^3) is response variable.

Table 8.7

x	y	x	y	x	y
10.40	10.20	15.10	14.80	16.50	15.90
10.50	10.40	15.10	15.10	16.70	16.60
11.90	11.60	15.10	14.50	17.10	16.70
12.10	11.90	15.70	15.70	17.10	16.70
13.80	13.50	15.80	15.20	17.80	17.60
15.00	14.50	16.00	15.80	18.40	18.30

(1) Find the least square regression line.

(2) Conduct a hypothesis test for the significance of the linear equation.

(3) What is the predicted volume if an animal's weight is 17.6 kg.

(4) What is the 95% prediction interval of the volume for an animal with 17.6 kg.

Solution:

(1)

$$n=18,\ \bar{x}=15.006,\bar{y}=14.722,\ \sum_{i=1}^{18} x_i=270.1,\ \sum_{i=1}^{18} y_i=265$$

$$\sum_{i=1}^{18} x_i^2=4\ 149.39,\ \sum_{i=1}^{18} y_i^2=3\ 996.14,\ \sum_{i=1}^{18} x_i y_i=4\ 071.71$$

$$S_{xy}=4\ 071.71-\frac{1}{18}\times 270.1\times 265=95.237\ 8,S_{xx}=4\ 149.39-\frac{1}{18}\times 270.1^2=96.389\ 4.$$

Therefore, $\hat{\beta}_1=\dfrac{S_{xy}}{S_{xx}}=\dfrac{95.237\ 8}{96.389\ 4}=0.988$, $\hat{\beta}_0=\bar{y}-\hat{\beta}_1\bar{x}=14.722-0.988\times 15.006=-0.104$, the regression equation is

$$\hat{y}=-0.104+0.988x.$$

(2) Method 1: At the significant level $\alpha=0.05$, we want to test

$$H_0:\beta_1=0,H_1:\beta_1\neq 0.$$

The test statistic is

$$T=\frac{\hat{\beta}_1}{\hat{\sigma}}\sqrt{S_{xx}}\sim t(n-2).$$

The rejection region is

$$|t|\geqslant t_{\alpha/2}(n-2)=t_{0.025}(16)=2.119\ 9.$$

$\hat{\sigma}=\sqrt{\dfrac{\text{SSE}}{n-2}}=\sqrt{\dfrac{0.661\ 2}{16}}=0.203\ 3$, $|t|=\left|\dfrac{0.988}{0.203\ 3}\times\sqrt{96.389\ 4}\right|=47.712\ 7>2.119\ 9$. Since the sample data fall in the rejection region, we reject H_0. β_1 is not zero. The regression line is significant.

In this case, the p-value is

$$p=P[\ |t(16)|\geqslant 47.712\ 7]=2.2\times 10^{-16}<0.05.$$

The p-value approach yields the same conclusion.

Method 2:

$$\text{SST}=S_{yy}=3\ 996.14-\frac{1}{18}\times 265^2=94.751\ 1,df=18-1=17$$

$$\text{SSR}=\hat{\beta}_1^2 S_{xx}=0.988^2\times 96.389\ 4=94.089\ 9,df=1$$

$$\text{SSE}=\text{SST}-\text{SSR}=0.661\ 2,df=16$$

$$F=\frac{\dfrac{94.089\ 9}{1}}{\dfrac{0.661\ 2}{16}}=2\ 276.828.$$

The calculations are reported in Table 8.8.

Table 8.8

Source	Sum of squares	Degrees of Freedom	Mean Square	F-value	p-value
Model	SSR = 94.089 9	1	94.089 9	2 276.828	2.2×10^{-16}
Error	SSE = 0.661 2	16	0.041 3	—	—
Total	SST = 94.751 1	17	—	—	—

At the level $\alpha=0.05$, $2\ 276.828>F_{0.05}(1,16)=4.494$, we reject H_0. The regression model is significant.

In this case, the p-value is the probability that we would observe an $F(1,16)$ random variable more extreme than 2 276.828,

$$p=P[F(1,16)\geqslant 2\ 276.828]=2.2\times10^{-16}<0.05.$$

The p-value approach yields the same conclusion.

(3) When $x_0=17.6$, the predicted volume of y_0 is $\hat{y}_0=\hat{\beta}_0+\hat{\beta}_1 x_0=-0.104+0.988\times17.6=17.284\ 8$.

(4) The 95% prediction interval of the volume for an animal with 17.6 kg is

$$\left(\hat{y}_0\pm t_{\alpha/2}(n-2)\hat{\sigma}\sqrt{1+\frac{1}{n}+\frac{(x_0-\bar{x})^2}{S_{xx}}}\right)$$

$$=\left(17.284\ 8\pm2.119\ 9\times0.203\ 3\sqrt{1+\frac{1}{18}+\frac{(17.6-15.006)^2}{96.389\ 4}}\right)$$

$$=(16.828,17.742).$$

See Figure 8.2.

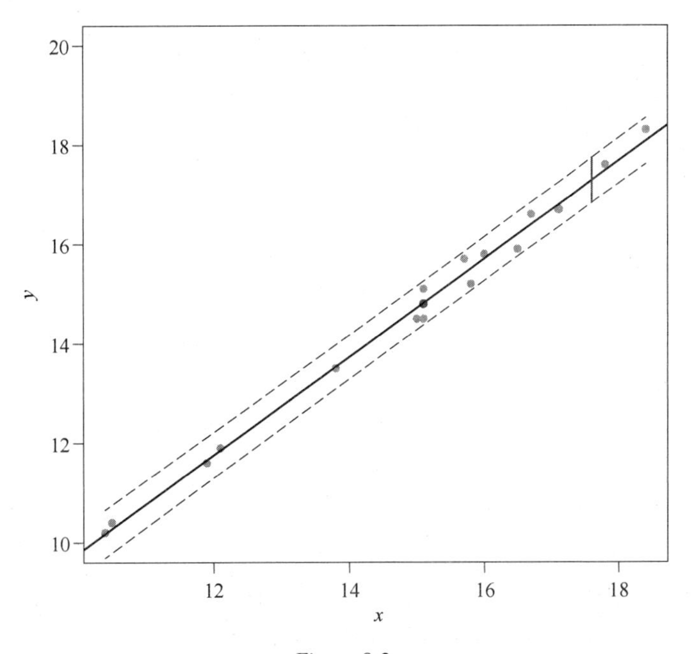

Figure 8.2

Bibliography

[1] 桂文豪，王立春，孔令臣．概率论与数理统计．2 版．北京：北京交通大学出版社，2023.

[2] 桂文豪，王立春．概率论与数理统计学习辅导及 R 语言解析．修订版．北京：北京交通大学出版社，2021.

[3] MORRIS H D, MARK J S. Probability and statistics. 4th ed. Boston: Addison-Wesley, 2018.

[4] JAY L D. Probability and statistics for engineering and the sciences. 9th ed. America: Cengage Learning, 2015.

[5] SHELDON R. A first course in probability. 10th ed. Princeton: Pearson Prentice Hall, 2020.